Volume 19 · 1988 · Number 6

optimization

a journal of mathematical programming and operations research

Formerly a series of

Mathematische
Operationsforschung
und Statistik

THEORIA CUM PRAXI

Akademie-Verlag · Berlin

6

ISSN 0233-1934 optimization, Berlin 19 (1988) 6, 761–908

Instructions for Authors

Submission of papers. Papers to be submitted for publication of books to be reviewed should be sent to

Professor K.-H. Elster
Editor-in-Chief, optimization
TH Ilmenau
Sektion Mathematik, Rechentechnik und ökonomische Kybernetik
Ehrenberg, Block G, DDR - 6300, Ilmenau,

Papers may be written in English (preferentially), French, German or Russian. It is assumed, that the papers have not been published previously. The original (or xerox copy) should be submitted with two additional copies. The manuscript will not normally be returned to the author. Authors are held fully responsible for all information included in their papers.

Preparation of manuscripts. Manuscripts should be typewritten on one side only, entirely double-spaced, including references, with wide margins on all four edges. Pages should be numbered consecutively, and the manuscripts should be arranged in the following order: article title, abbreviated title (to be used as running head, it should not exceed 35 characters), author's name and address; summary, AMS 1980 subject classification numbers and key words; main body of article; appendices; references; tables and figures. Before having typed a paper for submission, please examine a recent issue of *optimization* carefully and follow the arrangement of formulae, references, tables etc. Headings should be numbered by arabic numbers. The exact position of mathematical characters must be absolutely unambiguous. Distinguish between "oh" and "zero", "el" and "one", "kappa" and "kei", etc.

Text (with exception of mathematical symbols and formulae) may be stressed manually in the following ways:

Bold	straight black underline
Italics	waved blue underline
Spaced types	dotted underline
Small capitals	waved red underline

The following script annotation rules shall be observed for mathematical symbols and formulae:
Greek letters (excluding summation and product signs) should

be underlined		with a red line
bold roman types	(**A**, **B**, ...)	with a black line
bold italics	(***A***, ***B***, ...)	with two blue lines
gothic types	($\mathfrak{A}$, $\mathfrak{B}$, ...)	with a green line
bold gothic types	($\boldsymbol{\mathfrak{A}}$, $\boldsymbol{\mathfrak{B}}$, ...)	with two green lines
straight types	(A, B, ...)	with a violet line
script	($\mathcal{A}$, $\mathcal{B}$, ...)	with a brown line

Summary and key words. A selfcontained summary of 150 or fewer words should be submitted on a separate page with the manuscript. It should be followed by the American Mathematical Society 1980 subject classification numbers (under the heading "AMS 1980 Subject Classification: Primary —; Secondary —") and by three to ten key words indicating the contents of the article. The classification numbers representing the primary and secondary subjects of the article may be found with instructions for its use in the *Mathematical Reviews Annual Subject Index*–1980.

Tables and figures. Tables and figures should be titled, numbered consecutively, put on separate pages, and placed at the end of the manuscript. Appropriate placement in the text should be indicated in the margin of the manuscript. The number of illustrations should be kept as low as possible.

Footnotes. Footnotes should be reduced to a minimum and, where possible, should be replaced by remarks in the text; formulae in footnotes should be avoided. Footnotes, if allowed, cannot be cued next to a mathematical symbol, and should be numbered consecutively and typed together, double-spaced, on a separate page at the end of the manuscript.

Proofs and reprints. Authors will receive page proofs (in two sets). One corrected set of proofs should be returned to the editor within one week. Authors will be charged for alterations in excess of 10 % of the cost of composition.
75 reprints are offered free; up to 250 additional reprints are provided at cost price.

optimization

a journal of mathematical programming and operations research

Edited at
Sektion Mathematik, Rechentechnik und ökonomische Kybernetik
Technische Hochschule Ilmenau

Volume 19 · 1988

AKADEMIE-VERLAG · BERLIN

Editorial Policy

The journal publishes theoretical and applied papers related to the fields of linear, nonlinear, stochastic, parametric, discrete and dynamic programming, control theory, game theory, mathematical modelling of economical and technological processes, inventory, queueing theory, reliability, renewal theory, network methods etc. It is expected, that the papers give interesting and novel contributions to the theory at a high mathematical level or a stimulating presentation of ideas in foundations or applications. All papers should begin with a clear introduction including practical or theoretical consequences and connections to previous works. The trend to highest generality should not obscure the essential ideas.
A special section will be devoted to survey papers on theory and methods in interesting areas of mathematical programming and optimization techniques. Furthermore, the journal publishes proceedings of conferences, book reviews and announcements on related topics. Experts will be asked to give an opinion on each paper submitted for publication.

Terms of subscription

Orders can be sent

- in the GDR: to Postzeitungsvertrieb, or to Akademie-Verlag Berlin, Leipziger Str. 3–4, Pf-Nr. 1233, DDR - 1086 Berlin;
- in the other socialist countries: to a book-shop for foreign language literature or to the competent news-distributing agency;
- in the FRG and Berlin (West): to a book-shop or to the wholesale distributing agency Kunst und Wissen, Erich Bieber oHG, Postfach 102844, D - 7000 Stuttgart 10;
- in the other western european countries: to Kunst und Wissen, Erich Bieber GmbH, General Wille-Str. 4, CH - 8002 Zürich;
- in other countries: to the international book- and journalselling trade; to Buchexport, Volkseigener Außenhandelsbetrieb der DDR, Postfach 160, DDR - 7010 Leipzig; or to the Akademie-Verlag Berlin, Leipziger Str. 3–4, Pf-Nr. 1233, DDR - 1086 Berlin.

Zeitschrift „optimization"
Herausgeber: Akademie-Verlag Berlin, Leipziger Straße 3–4, DDR - 1086 Berlin; Fernruf: 2236221 oder 2236229; Telex-Nr.: 114420; Bank: Staatsbank der DDR, Berlin, Kto.-Nr.: 6836-26-27012.
Chefredakteur: Prof. Dr. Karl-Heinz Elster.
Verantwortlicher Redakteur: Dr. Axel Wolf.
Anschrift der Redaktion: Sektion Mathematik, Rechentechnik und ökonomische Kybernetik der TH Ilmenau, Postfach 327, DDR - 6300 Ilmenau.
Veröffentlicht unter der Lizenznummer 1586 des Presseamtes beim Vorsitzenden des Ministerrates der Deutschen Demokratischen Republik.
Gesamtherstellung: VEB Druckerei „G. W. Leibniz", DDR - 4450 Gräfenhainichen.
Erscheinungsweise: Die Zeitschrift „optimization" erscheint jährlich in einem Band mit 6 Heften. Bezugspreis je Band 198,– DM zuzüglich Versandspesen; Preis je Heft 33,– DM. Der gültige Jahresbezugspreis für die DDR ist der Postzeitungsliste zu entnehmen.
Bestellnummer dieses Bandes: 1093/19.

AN (EDV) 14829
02000

Contents of Volume 19 (1988)

Original papers

Authors

optimization

a journal of mathematical programming and operations research

Edited at
Sektion Mathematik, Rechentechnik und ökonomische Kybernetik
Technische Hochschule Ilmenau

Volume 19 · 1988 · Number 6

AKADEMIE-VERLAG · BERLIN

Editorial Policy

The journal publishes theoretical and applied papers related to the fields of linear, nonlinear, stochastic, parametric, discrete and dynamic programming, control theory, game theory, mathematical modelling of economical and technological processes, inventory, queueing theory, reliability, renewal theory, network methods etc. It is expected, that the papers give interesting and novel contributions to the theory at a high mathematical level or a stimulating presentation of ideas in foundations or applications. All papers should begin with a clear introduction including practical or theoretical consequences and connections to previous works. The trend to highest generality should not obscure the essential ideas.
A special section will be devoted to survey papers on theory and methods in interesting areas of mathematical programming and optimization techniques. Furthermore, the journal publishes proceedings of conferences, book reviews and announcements on related topics. Experts will be asked to give an opinion on each paper submitted for publication.

Terms of subscription

Orders can be sent
- in the GDR: to Postzeitungsvertrieb, or to Akademie-Verlag Berlin, Leipziger Str. 3–4, Pf-Nr. 1233, DDR - 1086 Berlin;
- in the other socialist countries: to a book-shop for foreign language literature or to the competent news-distributing agency;
- in the FRG and Berlin (West): to a book-shop or to the wholesale distributing agency Kunst und Wissen, Erich Bieber oHG, Postfach 102844, D - 7000 Stuttgart 10;
- in the other Western European countries: to Kunst und Wissen, Erich Bieber GmbH, General Wille-Str. 4, CH - 8002 Zürich;
- in other countries: to the international book- and journalselling trade; to Buchexport, Volkseigener Außenhandelsbetrieb der DDR, Postfach 160, DDR - 7010 Leipzig; or to the Akademie-Verlag Berlin, Leipziger Str. 3–4, Pf-Nr. 1233, DDR - 1086 Berlin.

Zeitschrift „optimization“
Herausgeber: Akademie-Verlag Berlin, Leipziger Straße 3–4, Berlin, DDR - 1086; Fernruf: 2236221 oder 2236229; Telex-Nr.: 114420; Bank: Staatsbank der DDR, Berlin, Kto.-Nr.: 6836-26-27012.
Chefredakteur: Prof. Dr. Karl-Heinz Elster.
Verantwortlicher Redakteur: Dr. Axel Wolf.
Anschrift der Redaktion: Sektion Mathematik, Rechentechnik und ökonomische Kybernetik der TH Ilmenau, Postfach 327, DDR - 6300 Ilmenau.
Veröffentlicht unter der Lizenznummer 1586 des Presseamtes beim Vorsitzenden des Ministerrates der Deutschen Demokratischen Republik.
Gesamtherstellung: VEB Druckerei „G. W. Leibniz“, DDR - 4450 Gräfenhainichen.
Erscheinungsweise: Die Zeitschrift „optimization“ erscheint jährlich in einem Band mit 6 Heften. Bezugspreis je Band 198,– DM zuzüglich Versandspesen; Preis je Heft 33,– DM. Der gültige Jahresbezugspreis für die DDR ist der Postzeitungsliste zu entnehmen.
Bestellnummer dieses Heftes: 1093/19/6.

AN (EDV) 14829
02000

optimization 19 (1988) 6, 763–780

A Covering Theorem for Set-Valued Mappings

A. Ya. KRUGER
PTO "Avtotranssystema" Minsk

Summary: The notion of covering is introduced for a set-valued mapping defined on an arbitrary set in a BANACH space. A necessary and sufficient covering criterion is proved. The conditions are formulated in terms of generalized differentials and generalized normals. The covering theorem is applied to deduce formulas of generalized differential calculus and necessary optimality conditions for nonsmooth optimization problems.

AMS 1980 Subject Classification:

Primary: 49 A 52; Secondary: 90 C 48, 58 C 06

Key words: Nonsmooth analysis, generalized differentials, set-valued mappings, optimality conditions, LAGRANGE multipliers.

1. Introduction

Rapid development of nonsmooth optimization in recent decade has led to numerous attempts to extend the main facts of classical analysis to nondifferentiable and nonconvex functions [4–9, 11, 16–21, 34, 35, 40–42, 44–52]. In its turn the formation of nonsmooth analysis apparatus affects the technique of proving optimality conditions in nonsmooth problems making it similar to the traditional schemes based on studying of some version of *covering property*.

In the simplest version this property for a mapping f at a point x^0 means that the image of every neighbourhood of x^0 under f contains a neighbourhood of $f(x^0)$ (cf. [11, 15, 51, 52]). Its role in optimization theory is motivated by the following evident observation: "every sufficient covering criterion is a source of necessary optimality conditions" [11].

We investigate here some different notion of covering: namely the *covering in a neighbourhood* [11] and extend it on set-valued mappings (section 3). This notion corresponds more precisely to the traditional analysis technique of deducing necessary conditions (in the continuously differentiable case covering at a point implies covering in a neighbourhood) and simplifies its extension on nonsmooth problems.

Several *covering constants* are introduced in sections 3 and 5 to characterize the covering property. Some modifications of *generalized differentials* [23–25, 28, 29, 32, 35] are used in the definitions. Replacement of CLARKE's *generalized gradients* [3, 6, 7, 46] (see covering constants in [11, 19]) by generalized differentials leads to necessary and sufficient *covering criteria* (theorems 3.1, 5.3 and the corollaries). The proof of the sufficient part is based on EKELAND's *variational principle* [13].

In the single-valued case the *covering theorem* (its statement is given in [27]) strenthens the covering criterion from [11]. We refer to [38] for some covering criterion in finite dimensions.

The second section deals with generalized differentials and generalized normals. The *lower* and the *upper* ε*-semidifferentials* (2.1) and (2.2) are introduced for a real-valued function defined on a set. Similar objects where used by many authers (see e.g. [13, 20, 21, 23, 24, 40, 46]) to characterize local properties of nonsmooth functions but their role in optimization theory was limited by rather poor calculus. Nevertheless it is possible to enrich the collection of properties of ε-semidifferentials by applying some perturbation process. The limiting objects (2.3), (2.4), namely *generalized semidifferentials* turn out to be very convenient for nonsmooth analysis and optimization leading to new results streng thening and extending the results of F. H. CLARKE, J. WARGA and other authors (see [20, 21, 25, 26, 28–33, 36, 37, 39]). We can strengthen them further if using in the statements the "perturbed" ε-semidifferentials themselves without taking limits, the results being in a certain sense the strongest possible. Some examples of this kind will be deduced from the covering theorem in sections 4 and 6.

The following assumptions are used in the paper. X, Y are real BANACH spaces and X^*, Y^* are their duals, $\|\cdot\|$, $\|\cdot\|$, $\langle\cdot,\cdot\rangle$ are accordingly the norms and the pairing between the spaces. $B_\varrho(z)$ denotes the closed ball of radius $\varrho>0$ with a center at z. The closing operation is denoted by a symbol *cl* (the closure of a set in a dual space is understood as a collection of all weak* limiting points of bounded sequences of its elements).

If not specified the norm in a product of a finite number of BANACH spaces is defined in an ordinary way: $\|u, v, \ldots\| = \max(\|u\|, \|v\|, \ldots)$. In $X\times Y$ the following norms will sometimes be used: $\|x, y\|_\alpha = \max(\|x\|, \alpha\|y\|)$, $x\in X$, $y\in Y$, where $\alpha>0$. If index α is not indicated it is supposed that $\alpha=1$.

Let us say that a norm of a BANACH space is *smooth* if it is FRECHET differentiable at all non-zero points. The collection of all BANACH spaces with a smooth norm will be denoted by $\mathfrak{X}$, and the collection of all BANACH spaces having an equivalent smooth norm will be denoted by $\mathfrak{Y}$. Let us note that $\mathfrak{Y}$ contains in particular all reflexive BANACH spaces [10].

2. Generalized Differentials and Generalized Normals

Let $\Omega\subset X$, $\varphi:\Omega\to\mathbb{R}$, $x\in\Omega$. The sets

$$\partial_\varepsilon^-\varphi(x) = \left\{x^*\in X^* \mid \liminf_{\substack{u\to x\\ u\in\Omega}} \frac{\varphi(u)-\varphi(x)-\langle x^*, u-x\rangle}{\|u-x\|} \geqq -\varepsilon\right\}, \tag{2.1}$$

$$\partial_\varepsilon^+\varphi(x)=\left\{x^*\in X^* \mid \liminf_{\substack{u\to x\\ u\in\Omega}} \frac{\varphi(u)-\varphi(x)-\langle x^*, u-x\rangle}{\|u-x\|}\leqq\varepsilon\right\} \tag{2.2}$$

will be called the *lower* and the *upper ε-semidifferentials* of φ at x (when $\varepsilon=0$ the index will be omitted).

The fact that a function φ is not allowed to take on infinite values does not restrict the applicability of the notions but makes it easier to extend them on mappings (set-valued in particular) taking values in BANACH spaces. The sets (2.1), (2.2) are evidently convex and are connected by the equality

$$\partial_\varepsilon^-(-\varphi)(x)=-\partial_\varepsilon^+\varphi(x)\,.$$

If $x\in\operatorname{int}\Omega$ and φ is FRÉCHET differentiable at x, then the semidifferentials $\partial^-\varphi(x)$, $\partial^+\varphi(x)$ are reduced to a derivative. If φ is convex (concave) then the lower semidifferential $\partial^-\varphi(x)$ (the upper semidifferential $\partial^+\varphi(x)$) coincides to a subdifferential (superdifferential) of φ at x in the sense of convex analysis [22, 42, 43]. When φ is convex the following equality holds: $\partial_\varepsilon^-\varphi(x)=\partial^-\varphi(x)+B_\varepsilon(0)$. Let us denote

$$\partial_\varepsilon^0\varphi(x)=\partial_\varepsilon^-\varphi(x)\cup\partial_\varepsilon^+\varphi(x)\,,$$
$$\hat\partial_\varepsilon\varphi(x)=\partial_\varepsilon^-\varphi(x)\cup\left(-\partial_\varepsilon^+\varphi(x)\right)\,.$$

On the basis of ε-semidifferentials (2.1), (2.3) the *lower* and the *upper generalized semidifferentials* of φ at x^0 can be defined:

$$\begin{aligned} D^-\varphi(x^0)&=\bigcap_{\varepsilon>0} cl\bigcup\{\partial_\varepsilon^-(cl^\downarrow\varphi)(x)\mid x\in cl\Omega\cap B_\varepsilon(x^0)\,,\\ &\qquad |cl^\downarrow\varphi(x)-\varphi(x^0)|\leqq\varepsilon\}\,,\\ D^+\varphi(x^0)&=\bigcap_{\varepsilon>0} cl\bigcup\{\partial_\varepsilon^+(cl^\uparrow\varphi)(x)\mid x\in cl\Omega\cap B_\varepsilon(x^0),\,|cl^\uparrow\varphi(x)-\varphi(x^0)|\leqq\varepsilon\}\,, \end{aligned}\tag{2.3}$$

The symbols $cl^\downarrow\varphi$ and $cl^\uparrow\varphi$ denote correspondingly the lower and the upper semicontinuous envelopes of φ:

$$cl^\downarrow\varphi(x)=\liminf_{\substack{u\to x\\ u\in\Omega}}\varphi(u),\quad cl^\uparrow\varphi(x)=\limsup_{\substack{u\to x\\ u\in\Omega}}\varphi(u),\quad x\in cl\Omega\,.$$

The generalized semidifferentials also generalize the notions of the subdifferential and the superdifferential. They reduce to a derivative if φ is strictly differentiable [2] at x^0. The sets (2.3), (2.4) may be nonconvex. They may be simultaneously non-empty and do not contain each other (see [25, 31, 32]). The need for two derivative-like objects to characterize local properties of functions is the intrinsic property of nonsmooth analysis (cf. [8, 9]). The reason is that the "differential" properties of φ and $-\varphi$ are essentially different in nonsmooth case. The following equality holds:

$$D^-(-\varphi)(x^0)=-D^+\varphi(x^0)\,. \tag{2.5}$$

The *generalized differential* and the *symmetrical generalized differential* of φ at x^0 are defined by the formulas:

$$D^0\varphi(x^0)=D^-\varphi(x^0)\cup D^+\varphi(x^0)\,, \tag{2.6}$$
$$\hat D\varphi(x^0)=D^-\varphi(x^0)\cup\left(-D^+\varphi(x^0)\right)\,. \tag{2.7}$$

Later on when speaking about the collection of the sets (2.3), (2.4), (2.6), (2.7) we will call them the generalized differentials.

If φ is continuous in a neighbourhood of x^0 and Ω is closed then the formulas (2.3), (2.4) can be simplified:

$$D^-\varphi(x^0) = \bigcap_{\varepsilon>0} cl \bigcup \{\partial_\varepsilon^- \varphi(x) \mid x \in \Omega \cap B_\varepsilon(x^0)\} ,$$

$$D^+\varphi(x^0) = \bigcap_{\varepsilon>0} cl \bigcup \{\partial_\varepsilon^+ \varphi(x) \mid x \in \Omega \cap B_\varepsilon(x^0)\} .$$

Unlike the ε-semidifferentials (2.1), (2.2) the lower ε-semidifferentials (2.3), (2.4) do not depend on the concrete norm of the space X.

The theorem below gives the relation between the generalized differential and the generalized gradient of CLARKE $\partial_C\varphi(x^0)$, the latter being defined [44–46] as a collection of all $x^* \in X^*$ such that $\langle x^*, \bar{x}\rangle \leqq \varphi^\dagger(x^0, \bar{x})$ for all $\bar{x} \in X$, where

$$\varphi^\dagger(x^0, \bar{x}) = \lim_{\varrho\to+0} \sup_{\substack{0<t<\varrho \\ (x,\mu)\in B_\varrho(x^0,\varphi(x^0)) \\ x\in\Omega,\varphi(x)\leqq\mu}} \inf_{\bar{u}\in B_\varrho(\bar{x})} \frac{\varphi(x+t\bar{u})-\mu}{t}$$

is the upper subderivative [44] of φ at x^0 in the direction $\bar{x}$ (the limit may be infinite).

Theorem 2.1: *For any $\varepsilon \geqq 0$, $\varrho \geqq 0$ and any weak* neighbourhood W of zero in X^* there exists an $\eta > 0$ such that*

$$\partial_\varepsilon^-(cl^\downarrow\varphi)(x) \cap B_\varrho(0) \subset \partial_C\varphi(x^0) + B_\varepsilon(0) + W$$

for all $x \in \Omega \cap B_\eta(x^0)$ such that $|cl^\downarrow\varphi(x) - \varphi(x^0)| \leqq \eta$.

Proof: Let $\bar{x} \in X$, $\alpha > 0$. By the definition of the upper subderivative there exists such an $\eta > 0$, that $\eta(\varrho+\varepsilon) < \frac{\alpha}{2}$ and for every $t \in (0, \eta)$, $(x, \mu) \in B_\eta(x^0, \varphi(x^0))$, $cl^\downarrow\varphi(x) \leqq \mu$, there could be indicated such an element $\bar{u}_t \in B_\eta(\bar{x})$ that

$$\frac{cl^\downarrow\varphi(x+t\bar{u}_t)-\mu}{t} < \varphi^\dagger(x^0; \bar{x}) + \frac{\alpha}{2} .$$

If also $x^* \in \partial_\varepsilon^-(cl^\downarrow\varphi)(x) \cap B_\varrho(0)$ then

$$\langle x^*, \bar{x}\rangle \leqq \liminf_{t\to+0} \frac{cl^\downarrow\varphi(x+t\bar{u}_t)-\mu}{t} + \varepsilon(\|\bar{x}\| + \eta) + \eta\varrho < \varphi^\dagger(x^0; \bar{x}) + \varepsilon\|\bar{x}\| + \alpha .$$

Hence, by the definition of the generalized gradient it follows that

$$\inf_{u^*\in\partial_C\varphi(x^0)+B_\varepsilon(0)} |\langle x^* - u^*, \bar{x}\rangle| < \alpha$$

and there exists an element $\hat{u}^* \in \partial_C\varphi(x^0) + B_\varepsilon(0)$ that $|\langle x^* - \hat{u}^*, \bar{x}\rangle| < \alpha$. This completes the proof of the theorem. ■

Corollary 2.1.1: (I) $D^-\varphi(x^0) \subset \partial_C\varphi(x^0)$.

(II) *If φ is* LIPSCHITZ *in a neighbourhood of x^0 then*

$$D^0\varphi(x^0) \subset \partial_C\varphi(x^0) .$$

The first statement is obvious. The second one follows the first by (2.5), (2.6) and the following property of the generalized gradient of CLARKE in the LIPSCHITZ case (see [5]):

$$\partial_C(-\varphi)(x^0) = -\partial_C\varphi(x^0) .$$

Corollary 2.1.1 shows that the generalized differentials being nonconvex turn out to be more acute local characteristics of functions than the generalized gradients of CLARKE. Besides this, it is possible to apply them to state the analogues of the classical differential calculus and convex functions, namely a subdifferential calculus (see [22, 42, 43]). According to A. D. IOFFE [21] the generalized differentials are the least (in the sense of inclusion) from all derivative generalizations possible with a certain collection of natural properties.

Let us define for Ω a *set of ε-normals*

$$N_\varepsilon(x \mid \Omega) = \left\{x^* \in X^* \mid \limsup_{\substack{u\to x\\ u\to\Omega}} \frac{\langle x\ , u-x\rangle}{\|u-x\|} \leqq \varepsilon\right\}, \quad x \in cl\Omega , \tag{2.8}$$

(when $\varepsilon = 0$ the index will be omitted) and a *cone of generalized normals*

$$K(x^0 \mid \Omega) = \bigcap_{\varepsilon>0} cl \cup \{N_\varepsilon(x \mid \Omega) \mid x \in cl\Omega \cap B_\varepsilon(x^0)\} . \tag{2.9}$$

In convex case the sets $N(x^0 \mid \Omega)$ and $K(y^0 \mid \Omega)$ reduce to a normal cone in the sense of convex analysis [22, 42, 43].

Let us consider a set-valued mapping $F: \Omega \to 2^Y$ with a graph $G(F) = \{(x, y) \in X \times Y \mid x \in \Omega,\ y \in F(x)\}$ and introduce a set

$$\partial_{\varepsilon,\alpha}F(x, y; y^*) = \{x^* \in X^* \mid (x^*, -y^*) \in N_{\varepsilon,\alpha}(x, y \mid G(F))\} , \tag{2.10}$$

where $(x, y) \in G(F)$, $y^* \in Y^*$, $\varepsilon \geqq 0$, $\alpha > 0$. Index α indicates that a norm $\|\cdot, \cdot\|_\alpha$ is used when defining a set of ε-normals to $G(F)$. The equality (2.10) defines a *(ε, α)-adjoint mapping* to F at a point (x, y). *Adjoint mapping* $DF(x^0, y^0; \cdot)$ to F at $(x^0, y^0) \in G(F)$ is defined similarly by the equality

$$DF(x^0, y^0; y^*) = \{x^* \in X^* \mid (x^*, -y^*) \in K(x^0, y^0 \mid G(F))\} .$$

In case of the mapping $(E\varphi)(x) = \varphi(x) + \mathbb{R}_+ = \{\mu \in \mathbb{R} \mid \varphi(x) \leqq \mu\}$, $x \in \Omega$, defined by a function $\varphi: \Omega \to \mathbb{R}$, for every $\varrho > 0$ there exists $\alpha > 0$ such that $\partial_{\varepsilon,\alpha}(E\varphi)(x, \varphi(x); 1) \cap B_\varrho(0) = \partial_\varepsilon^- \varphi(x) \cap B_\varrho(0)$. The following equalities hold:

$$\bigcap_{\alpha>0} \partial_{\varepsilon,\alpha}(E\varphi)\big(x, \varphi(x); 1\big) = \partial_\varepsilon^-\varphi(x) ,$$

$$D(E\varphi)\big(x^0, \varphi(x^0); 1\big) = D^-\varphi(x^0) .$$

For "horizontal" normals to the epigraph we set

$$\partial_{\varepsilon,\alpha}^\infty\varphi(x) = \partial_{\varepsilon,\alpha}(E\varphi)\big(x, \varphi(x); 0\big),\ D^\infty\varphi(x^0) = D(E\varphi)\big(x^0, \varphi(x^0); 0\big) .$$

We shall write $\partial_{\varepsilon,\alpha}f(x; y^*)$ and $Df(x^0; y^*)$ instead of $\partial_{\varepsilon,\alpha}F(x, f(x); y^*)$ and $DF(x^0, f(x^0); y^*)$ when $F(x) = \{f(x)\}$, $x \in \Omega$, $f: \Omega \to Y$. Let us set $\langle y^*, f\rangle(x) = \langle y^*, f(x)\rangle$, $x \in \Omega$. If f is LIPSCHITZ and $\alpha > 0$ is sufficiently small then $\partial_{\varepsilon,\alpha}f(x; y^*) = \partial_\varepsilon^-\langle y^*, f\rangle(x)$. If, besides this, dim $Y < \infty$ then $Df(x^0; y^*) = D^-\langle y^*, f\rangle(x^0)$.

Let us suppose that Ω is convex and a collection $\{\Lambda^{\varepsilon}f(x^0) \mid \varepsilon>0\}$ of nonvoid sets in the space of linear continuous functions from X into Y is a *directional derivate container* [50] of a continuous function $f: \Omega \to Y$ at x^0. A small modification of the proof of theorem 1.3 from [24] leads to the following result.

Theorem 2.2: *For all $\alpha>0$, $\varrho>0$, $y^* \in Y^*$ there exist $\varepsilon>0$, $\eta>0$ such that*

$$\partial_{\varepsilon}f(x;y^*) \subset y^*\Lambda^{\alpha}f(x^0)+N(x^0 \mid \Omega)+B_{\varrho}(0)$$

for every $x \in cl\Omega \cap B_{\eta}(x^0)$.

The following sets will be used in the next sections:

$$\delta_{\varepsilon,\alpha}F(x^0, y^0) = \bigcup\{\partial_{\varepsilon,\alpha}F(x,y;y^*) \mid x \in \Omega \cap B_{\alpha}(x^0), y \in F(x) \cap B_{\alpha}(y^0), \|y^*\|_* = 1\},$$

$$\delta_{\varepsilon,\alpha}f(x^0) = \bigcup\{\partial_{\varepsilon}^{-}\langle y^*, f\rangle (x) \mid x \in \Omega \cap B_{\alpha}(x^0), \|y^*\|_* = 1\},$$

$$\Delta F(x^0, y^0) = \bigcup\{DF(x^0, y^0; y^*) \mid \|y^*\|_* = 1\},$$

$$\Delta f(x^0) = \bigcup\{D^{-}\langle y^*, f\rangle (x^0) \mid \|y^*\|_* = 1\},$$

$$\delta_{\varepsilon,\alpha}^{-}\varphi(x^0) = \bigcup\{\partial_{\varepsilon}^{-}\varphi(x) \mid x \in \Omega \cap B_{\alpha}(x^0), \quad \varphi(x) \leqq \varphi(x^0)+\alpha\},$$

$$\delta_{\varepsilon,\alpha}^{\infty}\varphi(x^0) = \bigcup\{\partial_{\varepsilon,\alpha}^{\infty}\varphi(x) \mid x \in \Omega \cap B_{\alpha}(x^0), \quad \varphi(x) \leqq \varphi(x^0)+\alpha\}.$$

3. The Covering Theorem

Let Ω be a closed set in X and $F: \Omega \to 2^Y$ – an upper semicontinuous set-valued mapping (its graph $G(F)$ is closed in $X \times Y$), $(x^0, y^0) \in G(F)$, $a>0$. A mapping F *possesses a-covering property in a neighbourhood* of (x^0, y^0), if there exist neighbourhoods U of x^0 and V of y^0 such that for all $x \in \Omega \cap U$, $y \in F(x) \cap V$ we can find a number $r>0$ such that $F(\Omega \cap B_{\varrho}(x)) \supset B_{a\varrho}(y)$ for every $\varrho \in (0, r)$. Let us say that F *covers in a neighbourhood* of (x^0, y^0) if it possesses a-covering property in a neighbourhood of (x^0, y^0) for some $a>0$.

In case of ordinary (single-valued) continuous mapping $f: \Omega \to Y$ the definition of the covering property can be reformulated in the following manner.

A function f *possesses a-covering property in a neighbourhood of* $x^0 \in \Omega$ if there exists a neighbourhood U of x^0 such that for every $x \in \Omega \cap U$ we can find a number $r>0$ such that $f(\Omega \cap B_{\varrho}(x)) \supset B_{a\varrho}(f(x))$ for all $\varrho \in (0, r)$. *f covers in a neighbourhood* of x^0 if it possesses a-covering property in a neighbourhood of x^0 for some $a>0$.

Let us note that the covering property (contrary to the a-covering property for a fixed $a>0$) does not depend on the concrete norms in X and Y.

The following constant is used to characterize the covering property of a set-valued mapping F:

$$a_F(x^0, y^0) = \sup_{\alpha<0} \inf \{\varepsilon>0 \mid 0 \in \delta_{\varepsilon\alpha}F(x^0, y^0)\}. \tag{3.1}$$

Theorem 3.1: *Let $Y \in \mathfrak{X}$. The inequality $a \leqq a_F(x^0, y^0)$ is necessary for F to possess a-covering property in a neighbourhood of (x^0, y^0). The strict inequality $a < a_F(x^0, y^0)$ is sufficient.*

Proof: Let F possess a-covering property in a neighbourhood of (x^0, y^0) and $0 \in \partial_{\varepsilon\alpha} F(x, y; y^*)$ for some $\varepsilon > 0$, $\alpha > 0$, $x \in \Omega \cap B_\alpha(x^0)$, $y \in F(x) \cap B_\alpha(y^0)$, $\|y^*\|_* = 1$, that is

$$\sup_{\varrho>0} \inf_{\substack{(u,v)\in G(F) \\ \|(u,v)-(x,y)\|_\alpha \leqq \varrho}} \frac{\langle y^*, v - y\rangle}{\|(u, v) - (x, y)\|_\alpha} \geqq -\varepsilon \,. \tag{3.2}$$

It follows from the definition that $F(\Omega \cap B_\varrho(x)) \supset B_{a\varrho}(y)$ if the numbers $\alpha > 0$, $\varrho > 0$ are sufficiently small. If $\alpha a < 1$ then

$$\inf_{\substack{(u,v)\in G(F) \\ \|(u,v)-(x,y)\|_\alpha \leqq \varrho}} \frac{\langle y^*, v - y\rangle}{\|(u, v) - (x, y)\|_\alpha} \leqq \frac{1}{\varrho} \inf_{v \in B_{a\varrho}(y)} \langle y^*, v - y\rangle = -a \,. \tag{3.3}$$

Conditions (3.2), (3.3) imply the inequality $a \leqq a_F(x^0, y^0)$. ■

Let us suppose now that F does not possess a-covering property in a neighbourhood of (x^0, y^0), that is, for any neighbourhoods U of x^0 and V of y^0 we can find points $\hat{x} \in \Omega \cap U$, $\hat{y} \in F(\hat{x}) \cap V$, an arbitrarily small number $\varrho > 0$ and a point $\hat{v} \in B_{a\varrho}(\hat{y})$ such that $\|v - \hat{v}\| > 0$ for all $v \in F(\Omega \cap B_\varrho(\hat{x}))$. In accordance with EKELAND's theorem [13] for any $\alpha > 0$, $\varepsilon > a$ there exists a point $(x, y) \in G(F) \cap B_{a\varrho/\varepsilon}(\hat{x}, \hat{y})$ such that $0 < \|y - \hat{v}\| \leqq \|\hat{y} - \hat{v}\| \leqq a\varrho$ and $\|v - \hat{v}\| - \|y - \hat{v}\| + \varepsilon\| (u, v) - (x, y)\|_\alpha \geqq 0$ for all $u \in \Omega \cap B_\varrho(\hat{x})$, $v \in F(u)$. Hence $0 \in \partial_{\varepsilon,\alpha} F(x, y; y^*)$ where y^* is a derivative of the norm of Y at a point $y - \hat{v}$, and $a \geqq a_F(x^0, y^0)$.

Corollary 3.1.1: *Let $Y \in \mathfrak{Y}$. F covers in a neighbourhood of (x^0, y^0) if and only if $a_F(x^0, y^0) > 0$.*

The corollary is evident if we take into account the following simple fact. If $a_F^1(x^0, y^0)$ and $a_F^2(x^0, y^0)$ are the two covering constants corresponding to the two equivalent norms on Y then there exist numbers γ_1, γ_2 (not depending upon F, x^0, y^0) such that $\gamma_1 a_F^1(x^0, y^0) \leqq a_F^2(x^0, y^0) \leqq \gamma_2 a_F^1(x^0, y^0)$.

In case of a single-valued function $f\colon \Omega \to Y$ satisfying LIPSCHITZ condition in a neighbourhood of x^0 the constant (3.1) takes the following form:

$$a_f(x^0) = \sup_{\alpha>0} \inf \{\varepsilon > 0 \mid 0 \in \delta_{\varepsilon,\alpha} f(x^0)\} \,,$$

and the criterion of a-covering can be reformulated in the following manner.

Corollary 3.1.2: *Let $Y \in \mathfrak{X}$ and f is LIPSCHITZ in a neighbourhood of x^0. The inequality $a \leqq a_f(x^0)$ is necessary for f to possess a-covering property in a neighbourhood of x^0. The strict inequality $a < a_f(x^0)$ is sufficient.*

Corollary 3.1.3: *Let $Y \in \mathfrak{Y}$ and f is LIPSCHITZ in a neighbourhood of x^0. f covers in a neighbourhood of x^0 if and only if $a_f(x^0) > 0$.*

Remark: Instead of FRÉCHET differentiability of the norm in the statements above some other kind of differentiability can be used (e.g. in a GÂTEAUX sense) if we modify appropriately the definitions of ε-semidifferentials and ε-normals (cf. [20, 21]).

4. Applications to Calculus

The covering theorem is applied in this section to establish some facts of the ε-semidifferentials calculus.

Let us call a system of closed sets $\Omega_1, \Omega_2, \ldots, \Omega_n$ an X *extremal* [23, 26, 30] if $\bigcap_{i=1}^{n} \Omega_i \neq \emptyset$ and there exist sequences $\{a_k^i\} \subset X$, $i=1, 2, \ldots, n$, converging to 0 as $k \to \infty$ such that $\bigcap_{i=1}^{n} \Omega_{ik} = \emptyset$, where $\Omega_{ik} = \Omega_i - a_k^i = \{x - a_k^i \mid x \in \Omega_i\}$, $i=1, 2, \ldots, n$, $k=1, 2, \ldots$. Every point $x^0 \in \bigcap_{i=1}^{n} \Omega_i$ is called an extremal point of the system.

Theorem 4.1: *Let $X \in \mathfrak{Y}$. If x^0 is an extremal point of the system of closed sets Ω_1, $\Omega_2, \ldots, \Omega_n$ in X then for any $\varepsilon > 0$ there exist points $x_i \in \Omega_i \cap B_\varepsilon(x^0)$, $x_i^* \in N_\varepsilon(x_i \mid \Omega_i)$, $i=1, 2, \ldots, n$, such that* $\sum_{i=1}^{n} \|x_i^*\|_* = 1$ and

$$x_1^* + x_2^* + \ldots + x_n^* = 0 . \tag{4.1}$$

Proof: Let us consider a function $(u_1, u_2, \ldots, u_n) \to (u_n - u_1, u_n - u_2, \ldots, u_n - u_{n-1})$ from $\Omega_1 \times \Omega_2 \times \ldots \times \Omega_n$ into X^{n-1}. Evidently $X^{n-1} \in \mathfrak{Y}$. It follows from the definition of an extremal point that the function does not cover in a neighbourhood of a point $(x^0, x^0, \ldots, x^0) \in X^n$. Corollary 3.1.3 implies that for any $\varepsilon > 0$ there exist points $x_i \in \Omega_i \cap B_\varepsilon(x^0)$, $i=1, 2, \ldots, n$, $v^* = (v_1^*, v_2^*, \ldots, v_{n-1}^*) \in (X^{n-1})^*$ such that $\|v^*\|_* = 1$ and

$$\liminf_{\substack{u_i \to x_i \\ u_i \in \Omega_i \\ i=1,2,\ldots,n}} \frac{\sum_{i=1}^{n} \langle v_i^*, (u_n - u_i) - (x_n - x_i) \rangle}{\|(u_1, u_2, \ldots, u_n) - (x_1, x_2, \ldots, x_n)\|} \geqq -\varepsilon .$$

Let us set $v_n^* = -\sum_{i=1}^{n-1} v_i^*$, $x_i^* = v_i^* \Big/ \sum_{j=1}^{n} \|v_j^*\|_*$, $i=1, 2, \ldots, n$. It follows from the last inequality that $x_i^* \in N_\varepsilon(x_i \mid \Omega_i)$, $i=1, 2, \ldots, n$. Besides this, it is evident that $\sum_{i=1}^{n} \|x_i^*\|_* = 1$ and (4.1) holds. ■

Corollary 4.1.1: *Let the conditions of theorem 4.1 be satisfied and one of the following assumptions let be true:*

(1) *All the sets, except perhaps one, are epi-Lipschitzian at x^0.*

(2) dim $X < \infty$.

Then there exist points $x_i^* \in K(x^0 \mid \Omega_i)$, $i=1, 2, \ldots, n$, not all equal to 0 such that (4.1) *holds.*

These statements generalize the theorem of A. Ya. Dubovitskii-A. A. Milyutin [12] and the corresponding result of V. G. Boltyanskii [1]. Following [12] we shall call (4.1) the *generalized* Euler *equation.*

The next theorem gives the central result of the calculus of the generalized differentials – a formula for the lower ε-semidifferential of a sum of functions.

Theorem 4.2: *Let $X \in \mathfrak{Y}$, the sets Ω_1, Ω_2 let be closed in X, and the functions $\varphi_1: \Omega_1 \to \mathbb{R}$, $\varphi_2: \Omega_2 \to \mathbb{R}$ let be lower semicontinuous, $\Omega = \Omega_1 \cap \Omega_2$, $\varphi = \varphi_1 + \varphi_2: \Omega \to \mathbb{R}$. Let also $x^0 \in \Omega$ and we assume that there exists an $\eta > 0$ such that*

$$\delta^{\infty}_{\eta,\eta}\varphi_1(x^0) \cap \left(-\delta^{\infty}_{\eta,\eta}\varphi_2(x^0)\right) \subset B_1(0)\,.$$

Then the following inclusion holds:

$$\partial^-_\varepsilon \varphi(x) \subset \left(\delta^-_{\varepsilon_1,\varrho}\varphi_1(x) + \delta^-_{\varepsilon_2,\varrho}\varphi_2(x)\right) \tag{4.2}$$

for any $\varepsilon \geqq 0$, $\varepsilon_1 > 0$, $\varepsilon_2 > 0$, $\varepsilon_1 + \varepsilon_2 > \varepsilon$, $\varrho > 0$ and any $x \in \Omega$ such that $\|x - x^0\| < \eta$, $\varphi_i(x) < \varphi_i(x^0) + \eta$, $i = 1, 2$.

Proof: Let $x \in \Omega$, $\|x - x^0\| < \eta$, $\varphi_i(x) < \varphi_i(x^0) + \eta$, $i = 1, 2$, $\varepsilon \geqq 0$, $\varepsilon_1 > 0$, $\varepsilon_2 > 0$, $\varepsilon_1 + \varepsilon_2 > \varepsilon$, $\varrho > 0$ and $x^* \in \partial^-_\varepsilon \varphi(x)$. Let us consider an upper semicontinuous mapping $F: \Omega_1 \times \Omega_2 \to X \times 2^{\mathbb{R}}$ defined by the formula: $F(u_1, u_2) = \{(u_1 - u_2,\ \mu_1 + \mu_2 - \langle x^*, u_2 - x\rangle + \xi_1 \|u_1 - x\| + \xi_2 \|u_2 - x\|) \mid \varphi_1(u_1) \leqq \mu_1, \varphi_2(u_2) \leqq \mu_2\}$ where $0 \leqq \xi_1 < \varepsilon_1$, $0 \leqq \xi_2 < \varepsilon_2$, $\xi_1 + \xi_2 = \varepsilon$. It follows from (2.1) that F does not cover in a neighbourhood of $((x, x), (0, (\varphi)x)) \in G(F)$. Evidently $X \times \mathbb{R} \in \mathfrak{Y}$. Corollary 3.1.1 implies that $0 \in \delta_{\xi,\alpha} F((x, x), (0, \varphi(x)))$ for any $\xi > 0$, $\alpha > 0$. It follows from the last condition that there exist elements $x_1 \in \Omega_1$, $x_2 \in \Omega_2$, $\mu_1 \geqq \varphi_1(x_1)$, $\mu_2 \geqq \varphi_2(x_2)$, $u^* \in X^*$, $\lambda \in \mathbb{R}$ such that $\|x_1 - x\| \leqq \varrho$, $\|x_2 - x\| \leqq \varrho$, $|\eta_1 - \varphi_1(x)| \leqq \varrho$, $|\mu_2 - \varphi_2(x)| \leqq \varrho$, $\|u^*, \lambda\|_* = 1$ and

$$\liminf_{\substack{(u,\mu)\to(x_1,\mu_1)\\ u\in\Omega_1,\varphi_1(u)\leqq\mu}} \frac{\lambda\,(\mu - \mu_1) - \langle u^*, u - x_1\rangle}{\|u - x_1, \mu - \mu_1\|_\alpha} \geqq -(\xi + |\lambda|\xi_1)\,,$$

$$\liminf_{\substack{(u,\mu)\to(x_2,\mu_2)\\ u\in\Omega_2,\varphi_2(u)\leqq\mu}} \frac{\lambda\,(\mu - \mu_2) - \langle \lambda x^* - u^*, u - x_2\rangle}{\|u - x_2, \mu - \mu_2\|_\alpha} \geqq -(\xi + |\lambda|\xi_2)\,.$$

Without loss of generality we can suppose that $\alpha \leqq \eta$, $\xi \leqq \eta/4$ and $\|x_i - x^0\| \leqq \eta$, $\mu_i = \varphi_i(x_i) \leqq \varphi_i(x^0) + \eta$, $i = 1, 2$. Hence $u^* \in \partial^{\infty}_{\xi_1,\eta}\varphi_1(x_1)$, $-u^* \in \partial^{\infty}_{\xi_2,\eta}\varphi_2(x_2)$ where $\xi_1 = \xi + |\lambda|\ (\alpha^{-1} + \xi_1)$, $\xi_2 = \xi + |\lambda|\ (\alpha^{-1} + \xi_2 + \|x^*\|_*)$. Condition (1) implies that $|\lambda| \geqq \lambda_0 = \min\left(1/2,\ \frac{\eta}{4}\ (\varepsilon + \eta^{-1} + \|x^*\|_*)^{-1}\right)$, because otherwise we have $\xi_1 \leqq \frac{\eta}{2}$, $\xi_2 \leqq \frac{\eta}{2}$, $\|u^*\|_* \leqq 1/2$ and $\|u^*, \lambda\| < 1$. We have also $\lambda > -\alpha\xi$ and therefore $\lambda \geqq \lambda_0$ if α is sufficiently small. Let us set $x_1^* = u^*/\lambda$, $x_2^* = x^* - x_1^*$. Then $x_1^* \in \partial_{\eta_1,\alpha}(E\varphi_1)\ (x_1, \varphi_1(x_1); 1)$, $x_2^* \in \partial_{\eta_2,\alpha}(E\ \varphi_2)\ (x_2, \varphi_2(x_2); 1)$ where $\eta_i = \xi/\lambda + \xi_i$, $i = 1, 2$. Since $\|x_1^*\|_* \leqq \lambda_0^{-1}$, $\|x_2^*\|_* = \|x^*\|_* + \lambda_0^{-1}$ these conditions are equivalent to the inclusions $x_i^* \in \partial^-_{\eta_i}\ \varphi_i(x_i)$, $i = 1, 2$, if α is sufficiently small. We can evidently suppose that $\eta_1 \leqq \varepsilon_1$, $\eta_2 \leqq \varepsilon_2$. This completes the proof of (4.2). ■

Corollary 4.2.1: *Under the conditions of theorem 4.2 the following inclusion holds:*

$$D^-\varphi(x^0) \subset D^-\varphi_1(x^0) + D^-\varphi_2(x^0)\,. \tag{4.3}$$

The condition (1) is essential for the inclusions (4.2), (4.3) to be valid. They may be violated even in convex case. The condition holds for example if one of the functions φ_1, φ_2 is defined and LIPSCHITZ in a neighbourhood of x^0. A weaker condition for (1) to hold can be found in [25]. In finite dimensions condition (1) is equivalent to the following relation [38]:

(1′) $$D^{\infty}\varphi_1(x^0)\cap D^{\infty}\varphi_2(x^0)=\{0\}\,.$$

Corollary 4.2.2: *Let $X\in\mathfrak{Y}$, Ω_1, Ω_2 be closed sets in X, $\Omega=\Omega_1\cap\Omega_2$, $x^0\in\Omega$ and*

(2) *There exists an $\eta>0$ such that*

$$N_\eta(x_1\mid\Omega_1)\cap\big(-N_\eta(x_2\mid\Omega_2)\big)\subset B_1(0)$$

for any $x_1\in\Omega_1\cap B_\eta(x^0)$, $x_2\in\Omega_2\cap B_\eta(x^0)$.
Then the following inclusion holds:

$$N_\varepsilon(x\mid\Omega)\subset\bigcup\{N_{\varepsilon_1}(x_1\mid\Omega_1)+N_{\varepsilon_2}(x_2\mid\Omega_2)\mid x_i\in\Omega_i\cap B_\varrho(x),\ i=1,2\}$$

for any $\varepsilon\geqq 0$, $\varepsilon_1>0$, $\varepsilon_2>0$, $\varepsilon_1+\varepsilon_2>\varepsilon$, $\varrho>0$ and any $x\in\Omega$, $\|x-x^0\|<\eta$.

Corollary 4.2.3: *Under the conditions of corollary 4.2.2 the following inclusion holds:*

$$K(x^0\mid\Omega)\subset K(x^0\mid\Omega_1)+K(x^0\mid\Omega_2)\,.$$

If $\dim X<\infty$ condition (2) is equivalent to the equality

(2′) $$K(x^0\mid\Omega_1)\cap\big(-K(x^0\mid\Omega_2)\big)=\{0\}\,.$$

Some other facts of the calculus of generalized differentials are stated below without proofs. They can be deduced from theorem 4.2 by standard methods (see [29]). To simplify the statements all the functions are supposed to be locally LIPSCHITZ.

Theorem 4.3: *Let $X\in\mathfrak{Y}$, $\varphi_1\colon X\to\mathbb{R}$, $\varphi_2\colon X\to\mathbb{R}$, $\varphi=\max(\varphi_1,\varphi_2)$. Then the following conditions hold for any $x\in X$ such that $\varphi_1(x)=\varphi_2(x)$:*

(i) $$\partial_\varepsilon^-\varphi(x)\subset\bigcup\{\delta_{\varepsilon_1,\varrho}^-(\lambda_1\varphi_1)(x)+\delta_{\varepsilon_2,\varrho}^-(\lambda_2\varphi_2)(x)\mid\lambda_1\geqq 0,\ \lambda_2\geqq 0,\ \lambda_1+\lambda_2=1\}$$
for any
$\varepsilon\geqq 0$, $\varepsilon_1>0$, $\varepsilon_2>0$, $\varepsilon_1+\varepsilon_2>\varepsilon$, $\varrho>0$;

(ii) $$D^-\varphi(x)\subset\bigcup\{\lambda_1D^-\varphi_1(x)+\lambda_2D^-\varphi_2(x)\mid\lambda_1\geqq 0,\ \lambda_2\geqq 0,\ \lambda_1+\lambda_2=1\}\,.$$

Theorem 4.4: *Let X, $Y\in\mathfrak{Y}$, $f\colon X\to Y$, $\psi\colon Y\to\mathbb{R}$, $\varphi(x)=\psi(f(x))$, $x\in X$. Then the following conditions hold for any $x\in X$:*

(i) $$\partial_\varepsilon^-\varphi(x)\subset\bigcup\{\delta_{\eta,\varrho}^-\langle y^*,f\rangle(x)\mid y^*\in\delta_{\zeta,\varrho}^-\psi(f(x))\}$$
for any
$\varepsilon\geqq 0$, $\eta>\varepsilon$, $\zeta>0$, $\varrho>0$;

(ii) $$D^-\varphi(x)\subset\bigcup\{D^-\langle y^*,f\rangle(x)\mid y^*\in D^-\psi(f(x))\}\,.$$

Corollary 4.4.1: *Let the hypotheses of theorem 4.4 be satisfied and f be strictly differentiable at x^0. Then*

(i) *for any $\alpha>0$ there exists $\beta>0$ such that*

$$\partial_\varepsilon^-\varphi(x)\subset\big(\nabla f(x^0)\big)^*\,\delta_{\xi,\varrho}^-\psi\big(f(x)\big)+B_{\varepsilon+\alpha}(0)$$

if $\varepsilon>0$, $\xi>0$, $\varrho>0$, $x\in X\cap B_\beta(x^0)$;

(ii) $$D^-\varphi(x^0)\subset\big(\nabla f(x^0)\big)^*D^-\psi\big(f(x^0)\big)\,.$$

Corollary 4.4.2: *Let the hypotheses of theorem 4.4 be satisfied and ψ be strictly* *differentiable at* $f(x^0)$. *Then*

(i) *for any* $\alpha>0$ *there exists* $\beta>0$ *such that*

$$\partial_\varepsilon^- \varphi(x) \subset \delta_{\varepsilon+\alpha,\varrho}^- \langle \psi'(f(x^0)), f\rangle(x)$$

if $\varepsilon>0$, $\varrho>0$, $x\in X\cap B\,(x^0)$;

(ii) $$D^-\varphi(x^0)\subset D^-\langle \psi'(f(x^0)), f\rangle(x^0)\ .$$

Theorem 4.5: *Let* $X\in\mathfrak{Y}$, $\varphi\colon X\to\mathbb{R}$. *Then for any* $x_1, x_2\in X$ *and any* $\varepsilon>0$, $\varrho>0$ *there exist* $x^0\in\,]x_1, x_2[\,=\{x_1+t(x_2-x_1)\ 0<t<1\}$ *and* $x^*\in\delta_{\varepsilon,\varrho}^-\varphi(x^0)$ *such that* $\varphi(x_2)-\varphi(x_1)=\langle x^*, x_2-x_1\rangle$.

5. Other Covering Criteria

According to theorem 3.1 a question of an upper semicontinuous set-valued mapping F possessing a covering property in a neighbourhood of $(x^0, y^0)\in G(F)$ can be reduced to calculating the covering constant (3.1). The following constants can also be used for that purpose:

$$b_F(x^0, y^0)=\sup_{\alpha>0}\inf\{\|x^*\|_* \mid x^*\in\delta_{\alpha,\alpha}F(x^0, y^0)\}\ , \tag{5.1}$$

$$c_F(x, y)=\inf\{\|x^*\|_* \mid x^*\in\Delta F(x^0, y^0)\}\ . \tag{5.2}$$

Theorem 5.1: (i) $b_F(x^0, y^0)\geqq a_F(x^0, y^0)$.
(ii) $b_F(x^0, y^0)>0$ *if and only if* $a_F(x^0, y^0)>0$.
(iii) *If* $X, Y\in\mathfrak{Y}$ *then* $b_F(x^0, y^0)=a_F(x^0, y^0)$.

Proof: To prove (i) it is sufficient to notice that inclusion $x^*\in\partial_{\alpha,\alpha}F(x, y; y^*)$ implies inclusion $0\in\partial_{\varepsilon,\alpha}F(x, y; y^*)$ where $\varepsilon=\alpha+\|x^*\|^*$. (ii) It is evident from the definitions (3.1), (5.1) that each of the equalities $a_F(x^0, y^0)=0$ and $b_F(x^0, y^0)=0$ is equivalent to the following condition: $0\in\delta_{\alpha,\alpha}F(x^0, y^0)$ for any $\alpha>0$.

Now we shall prove (iii). Let $\varepsilon>a_F(x^0, y^0)$, $\alpha>0$. Then there exist elements $x\in\Omega\cap B_{\alpha/2}(x^0)$, $y\in F(x)\cap B_{\alpha/2}(y^0)$, $y^*\in Y^*$ such that $\|y^*\|_*=1$ and $0\in\partial_{\varepsilon,\alpha}F(x, y; y^*)$. The latter condition is equivalent to the inclusion $(0, -y^*)\in\partial^-(\delta(\cdot\mid G(F))+\varepsilon\|\cdot\|_\alpha)\,(x, y)$ ($X\times Y$ is considered with the norm $\|\cdot,\cdot\|_\alpha$) where $\delta(\cdot\mid G(F))\colon G(F)\to\mathbb{R}$ is an indicator function of $G(F)$ $(\delta(u, v\mid G(F))\equiv 0)$. If $X, Y\in\mathfrak{Y}$ then $X\times Y\in\mathfrak{Y}$. It follows from theorem 4.2 that for any $\xi>0$ there exist elements $(\hat{x}, \hat{y})\in G(F)\cap B(x, y)$, $u^*\in X^*$, $v^*\in Y^*$ such that $u^*\in\partial_{\xi,\alpha}(\hat{x}, \hat{y}; v^*)$, $\|u^*, v^*-y^*\|_{\alpha^*}\leqq\varepsilon+\xi$. Hence $\|u^*\|_*\leqq\varepsilon+\xi$, $\|v^*\|_*\geqq 1-\alpha(\varepsilon+\zeta)$. We have $\hat{x}\in\Omega\cap B_\alpha(x^0)$, $\hat{y}\in F(\hat{x})\cap B_\alpha(y^0)$, $\alpha(\varepsilon+\zeta)<1$ if α and ζ are sufficiently small. Let us set $\hat{x}^*=u^*/\|v^*\|_*$, $\hat{y}^*=v^*/\|v^*\|_*$. Then $\|\hat{y}^*\|_*=1$, $\|\hat{x}^*\|_*\leqq\varepsilon_1$, $\hat{x}^*\in\partial_{\varepsilon_2,\alpha}F(\hat{x}, \hat{y}; \hat{x}^*)$ where $\varepsilon_1=(\varepsilon+\xi)/(1-\alpha(\varepsilon+\xi))$, $\varepsilon_2=\xi/(1-\alpha\,(\varepsilon+\xi))$. Since $\varepsilon_1\to\varepsilon$, $\varepsilon_2\to 0$ as $\alpha\to 0$, $\zeta\to 0$, it follows that $b_F(x^0, y^0)\leqq\varepsilon$. This completes the proof. ∎

Let us say that a mapping $F\colon\Omega\to 2^Y$ is *uniformly epi-Lipschitzian* in a neighbourhood of $(x^0, y^0)\in G(F)$ if there exist neighbourhoods U of x^0 and V of y^0, an open set $W\subset Y$ and a number $\eta>0$ such that $y+t\bar{y}\in F(x)$ for all $x\in\Omega\cap U$, $y\in\ni F(x)\cap V$, $\bar{y}\in W$, $0<t<\eta$. If $F(x)=f(x)-M$ where M is a closed set in Y, f is a

LIPSCHITZ function from X into Y and $f(x^0) \in M$, then F is uniformly epi-Lipschitzian in a neighbourhood of $(x^0, 0)$ if and only if M is epi-Lipschitzian [44] in a neighbourhood of $f(x^0)$.

Theorem 5.2: (i) *If* $\dim Y < \infty$ *then* $c_F(x^0, y^0) \leqq b_F(x^0, y^0)$.

(ii) *If* $\dim X < \infty$, $\dim Y < \infty$ *then* $c_F(x^0, y^0) = b_F(x^0, y^0)$.

(iii) *If* $Y = Y_1 \times Y_2$, $\dim Y_2 < \infty$, $F = (F_1, F_2)$: $\Omega \to 2^{Y_1} \times 2^{Y_2}$, $y^0 = (y_1^0, y_2^0) \in Y_1 \times Y_2$ *and* F_1 *is uniformly epi-Lipschitzian in a neighbourhood of* (x^0, y_1^0) *then the equality* $b_F(x^0, y^0) = 0$ *implies that* $0 \in \Delta F(x^0, y^0)$.

Under the conditions of (i) *or* (*ii*) *an infimum in* (5.2) *is attained.*

Proof: (i) Let $\dim Y < \infty$. Then $\Delta F(x^0, y^0)$ is weakly* closed and an infimum in (5.2) is attained. In accordance with (5.1) there exist sequences $\{x_k, y_k\} \subset G(F)$, $\{x_k^*\} \subset X^*$, $\{y_k^*\} \subset Y^*$, $\{\alpha_k\} \subset \mathbb{R}_+$ such that $x_k^* \in \partial_{\alpha_k, \alpha_k} F(x_k, y_k; y_k^*)$, $\|y_k^*\|_* = 1$, $k = 1, 2, \ldots$, $x_k \to x^0$, $y_k \to y^0$, $\alpha_k \to 0$ and $\|x_k^*\|_* \to b_F(x^0, y^0)$ as $k \to \infty$. $\{x_k^*\}$, $\{y_k^*\}$ have weak* limiting points x^*, y^*. Evidently $x^* \in DF(x^0, y^0; y^*)$, $\|y^*\|_* = 1$ and $c_F(x^0, y^0) \leqq \|x^*\|_* \leqq b_F(x^0, y^0)$.

(ii) If $\dim X < \infty$, $\dim Y < \infty$ then $c_F(x, y) = \|x^*\|_*$ for some $x^* \in DF(x^0, y^0; y^*)$, $y^* \in Y^*$, $\|y^*\|_* = 1$. In accordance with the definition of a conjugate mapping for any α, $0 < \alpha < 1$, there exist elements $x_\alpha \in \Omega \cap B_\alpha(x^0)$, $y_\alpha \in F(x)_\alpha \cap B_\alpha(y^0)$, $u_\alpha^* \in B_{\alpha/(1+\alpha)}(x^*)$, $v_\alpha^* \in B_{\alpha/(1+\alpha)}(y^*)$ such that $u_\alpha^* \in \partial_{\alpha/(1+\alpha), \alpha} F(x_\alpha, y_\alpha; v_\alpha^*)$. We have also $\|u_\alpha^*\|_* \leqq \|x^*\|_* + \alpha/(1+\alpha)$, $\|v_\alpha^*\|_* \geqq 1/(1+\alpha)$. Let us set $x_\alpha^* = u_\alpha^*/\|v_\alpha^*\|_*$, $y_\alpha^* = v_\alpha^*/\|v_\alpha^*\|_*$. Then $\|y_\alpha^*\|_* = 1$, $\|x_\alpha^*\|_* \leqq \|x^*\|_* + \alpha(1 + \|x^*\|_*)$, $x_\alpha^* \in \partial_{\alpha,\alpha} F(x_\alpha, y_\alpha; y_\alpha^*)$ and consequently $b_F(x^0, y^0) \leqq c_F(x^0, y^0)$.

(iii) Let $Y = Y_1 \times Y_2$, $y^0 = (y_1^0, y_2^0) \in Y_1 \times Y_2$, $F = (F_1, F_2)$: $\Omega \to 2^{Y_1} \times 2^{Y_2}$, $b_F(x^0, y^0) = 0$. According to the definition (5.1) there exist sequences $\{x_k, (y_{1k}, y_{2k})\} \subset G(F)$, $\{x_k^*\} \subset X^*$, $\{y_{1k}^*\} \subset Y_1^*$, $\{y_{2k}^*\} \subset Y_2^*$, $\{\alpha_k\} \subset \mathbb{R}_+$ such that $x_k^* \in \partial_{\alpha_k, \alpha_k} F\big(x_k, (y_{1k}, y_{2k}); (y_{1k}^*, y_{2k}^*)\big)$, $\|y_{1k}^*, y_{2k}^*\|_* = 1$, $k = 1, 2, \ldots$, $x_k \to x^0$, $y_{1k} \to y_1^0$, $y_{2k} \to y_2^0$, $\alpha_k \to 0$ and $\|x_k^*\|_* \to 0$ as $k \to \infty$. If F_1 is uniformly epi-Lipschitzian in a neighbourhood of (x^0, y_1^0) then there exist a point $\bar{v} \in Y_1$ and a number $\eta > 0$ such that $y + tv \in F_1(x)$ for all $x \in \Omega \cap B_\eta(x^0)$, $y \in F_1(x) \cap B_\eta(y_1^0)$, $v \in B_\eta(\bar{v})$, $0 < t < \eta$. Without loss of generality we can suppose that $\{x_k\} \subset B_\eta(x^0)$, $\{y_{1k}\} \subset B_\eta(y_1^0)$ and the sequences $\{y_{1k}^*\}$, $\{y_{2k}^*\}$ have weak* limiting points y_1^*, y_2^*. We evidently have $0 \in DF\big(x^0, y^0; (y_1^*, y_2^*)\big)$, $\|y_1^*, y_2^*\|_* \leqq 1$. For every k there exists a point $v_k \in B_\eta(\bar{v})$ such that $\langle y_{1k}^*, v_k - \bar{v} \rangle \leqq -\|y_{1k}^*\|_* \eta + \alpha_k$. On the other hand since $y_{1k} + tv_k \in F_1(x_k)$ as $0 < t < \eta$, we have $-\langle y_{1k}^*, v_k \rangle \leqq \alpha_k \|v_k\|$. Hence $\langle y_{1k}^*, \bar{v} \rangle + \eta \, \|y_{2k}^*\|_* \geqq \eta - \alpha_k(1 + \|v_k\|)$. If $\dim Y_2 < \infty$ then we can suppose that $\|y_{2k}^*\|_* \to \|y_2^*\|_*$ as $k \to \infty$. Then $\langle y_1^*, \bar{v} \rangle + \eta \, \|y_2^*\|_* \geqq \eta$ and consequently $(y_1^*, y_2^*) \neq 0$, $0 \in \Delta F(x^0, y^0)$. ■

Now we can state several covering criteria for a mapping F.

Theorem 5.3: (i) *If* $Y \in \mathfrak{Y}$ *then* F *covers in a neighbourhood of* (x^0, y^0) *if and only if* $b_F(x^0, y^0) > 0$.

(ii) *If* $\dim X < \infty$, $\dim Y < \infty$ *then* F *covers in a neighbourhood of* (x^0, y^0) *if and only if* $0 \notin \Delta F(x^0, y^0)$.

(iii) *If* $Y = Y_1 \times Y_2$, $Y_1 \in \mathfrak{Y}$, $\dim Y_2 < \infty$, $F = (F_1, F_2)$: $\Omega \to 2^{Y_1} \times 2^{Y_2}$, $y^0 =$

$=(y_1^0, y_2^0)\in Y_1\times Y_2$, F_1 *is uniformly epi-Lipschitzian in a neighbourhood of* (x^0, y_1^0) *and* $0\notin \Delta F(x^0, y^0)$ *then* F *covers in a neighbourhood of* (x^0, y^0).

It must be noted that the cases dim $Y_1=0$ (mapping into a finite-dimensional space) and dim $Y_2=0$ (uniformly epi-Lipschitzian mapping) are not excluded from the statement of part (iii) of the theorem 5.3.

If $F(x)=\{f(x)\}$, $x\in\Omega$, and a function $f: \Omega\to Y$ is LIPSCHITZ in a neighbourhood of x^0 then the covering constants (5.1) and (5.2) take the following form:

$$b_f(x^0)=\sup_{\alpha>0}\inf\{\|x^*\|_* \mid x^*\in\delta_{\alpha,\alpha}f(x^0)\}\,, \tag{5.3}$$

$$c_f(x^0)=\inf\{\|x^*\|_* \mid x^*\in\Delta f(x^0)\}\,. \tag{5.4}$$

The theorems 5.1–5.3 imply the following results.

Theorem 5.4: (i) $b_f(x^0)\geqq a_f(x^0)$.
(ii) $b_f(x^0)>0$ *if and only if* $a_f(x^0)>0$.
(iii) *If* $X, Y\in\mathfrak{Y}$ *then* $b_f(x^0)=a_f(x^0)$.
(iv) *If* dim $Y<\infty$ *then* $c_f(x^0)\leqq b_f(x^0)$.
(v) *If* dim $X<\infty$, dim $Y<\infty$ *then* $c_f(x^0)=b_f(x^0)=a_f(x^0)$.
Under the conditions of (iv) *or* (v) *the infimum in* (5.4) *is attained.*

Theorem 5.5: (i) *Let* $Y\in\mathfrak{Y}$. f *covers in a neighbourhood of* x^0 *if and only if* $b_f(x^0)>0$.
(ii) *Let* dim $X<\infty$, dim $Y<\infty$. f *covers in a neighbourhood of* x^0 *if and only if* $0\notin\Delta f(x^0)$.
(iii) *If* dim $Y<\infty$, $0\notin\Delta f(x^0)$ *then* f *covers in a neighbourhood of* x^0.

It follows also from the theorems 5.1, 5.4 that in the case $X, Y\in\mathfrak{X}$ the constants (5.1), (5.3) can be used to characterize a-covering property.

6. Necessary Conditions for an Extremum

Let us consider two closed sets $\Omega\subset X$, $M\subset Y$ and a function $f: \Omega\to Y$ which is assumed to be LIPSCHITZ in a neighbourhood of a point $x^0\in\Omega$. We shall say that x^0 is (f, Ω, M)*-extremal* if $f(x^0)\in M$ and for any neighbourhoods U of x^0 in X and V of zero in Y and any $a>0$ there exist points $\hat{x}\in\Omega\cap U$, $\hat{y}\in(f(\hat{x})-M)\cap V$ and sequences $\{\varrho_k\}\subset\mathbb{R}_+$, $\{y_k\}\subset Y$ such that $\varrho_k\to 0$ as $k\to\infty$, $y_k\in B_{a\varrho_k}(\hat{y})$ and $\{u\in\Omega\cap B_{\varrho_k}(\hat{x}) \mid f(u)-y_k\in M\}=\emptyset$, $k=1, 2, \ldots$

The notion of "extremality" introduced above embraces various notions of a solution in optimization problems. The definition is a simple restatement of a requirement for a mapping $F(x)=f(x)-M$, $x\in\Omega$, not to cover in a neighbourhood of $(x^0, 0)\in G(F)$. To apply some covering criterion we need expressions for the sets $\delta_{\varepsilon,\alpha}F(x^0, y^0)$ and $\Delta F(x^0, y^0)$. Let us introduce some notations.

$$\delta^M_{\varepsilon,\alpha}f(x^0, y^0)=\cup\{\partial^-_\varepsilon\langle y^*, f\rangle(x) \mid y^*\in N_\alpha(y\mid M),\ \|y^*\|_*=1\,,\ x\in\Omega\cap B_\alpha(x^0)\,,\ y\in M\cap B_\alpha(y^0)\}\,,$$

$$\tilde{\delta}^M_{\varepsilon,\alpha}f(x^0, y^0)=\cup\{\partial^-_\varepsilon\langle y^*, f\rangle(x) \mid y^*\in N(y\mid M)\,,\ \|y^*\|_*=1\,,\ x\in\Omega\cap B_\alpha(x^0)\,,\ y\in M\cap B_\alpha(y^0)\}\,,$$

$$\delta^M_{\varepsilon,\alpha}f(x^0, y^0) = \cup\{\partial^-_\varepsilon\langle y^*, f\rangle(x) \mid y^* \in N(y^0 \mid M),\quad \|y^*\|_* = 1,\quad x \in \Omega \cap B(x^0)\},$$

$$\Delta^M f(x^0, y^0) = \cup\{D^-\langle y^*, f\rangle(x^0) \mid y^* \in K(y^0 \mid M),\quad \|y^*\|_* = 1\}.$$

We evidently have $\delta^M_{\varepsilon,\alpha}f(x^0, y^0) \subset \tilde{\delta}^M_{\varepsilon,\alpha}f(x^0, y^0) \subset \delta^M_{\varepsilon,\alpha}f(x^0, y^0) \subset \delta_{\varepsilon,\alpha}f(x^0)$ and $\Delta^M f(x^0, y^0) \subset \subset \Delta f(x^0)$. The theorem below is a simple consequence of the definitions.

Theorem 6.1: *Let $F(x) = f(x) - M$, $x \in \Omega$.*

(i) *For any $\varepsilon > 0$, $\alpha > 0$, $\xi \in (0, \varepsilon)$ there exists $\varrho \in (0, \alpha)$ such that* $\delta^M_{\xi,\varrho}f(x^0, y^0) \subset \subset \delta_{\varepsilon,\alpha}F(x^0, f(x^0) - y^0)$, $\delta_{\varepsilon,\varrho}F(x^0, f(x^0) - y^0) \subset \delta^M_{\varepsilon,\alpha}f(x^0, y^0)$.

(ii) $\Delta F(x^0, f(x^0) - y^0) = \Delta^M f(x^0, y^0)$.

When M is convex part (i) of theorem 6.1 can be strengthened.

Corollary 6.1.1: *Let $F(x) = f(x) - M$, $x \in \Omega$, and M is convex.*

(i) *For any $\varepsilon > 0$, $\alpha > 0$, $\xi \in (0, \varepsilon)$ there exists $\varrho > 0$ such that* $\tilde{\delta}^M_{\xi,\varrho}f(x^0, y^0) \subset \delta_{\varepsilon,\alpha}F(x^0, f(x^0) - y^0)$, $\delta_{\xi,\varrho}F(x^0, f(x) - y^0) \subset \tilde{\delta}^M_{\varepsilon,\alpha}f(x^0, y^0)$.

(ii) *If* $\dim Y < \infty$ *then for any $\varepsilon > 0$, $\alpha > 0$, $\xi \in (0, \varepsilon)$ there exists $\varrho > 0$ such that* $\delta^M_{\xi,\varrho}f(x^0, y^0) \subset \delta_{\varepsilon,\alpha}F(x^0, f(x^0) - y^0)$, $\delta_{\xi,\varrho}F(x^0, f(x^0) - y^0) \subset \delta^M_{\varepsilon,\alpha}f(x^0, y^0)$.

Application of theorems 5.3 and 6.1 leads to the following extremality conditions.

Theorem 6.2: (i) *Let $Y \in \mathfrak{Y}$. x^0 is (f, Ω, M)-extremal if and only if for any $\varepsilon > 0$ there exist elements $x \in \Omega \cap B_\varepsilon(x^0)$, $y \in M \cap B_\varepsilon(f(x^0))$, $y^* \in Y^*$, $\|y^*\|_* = 1$, such that*

$$y^* \in N_\varepsilon(y \mid M), \tag{6.1}$$

$$0 \in \partial^-_\varepsilon\langle y^*, f\rangle(x). \tag{6.2}$$

(ii) *Let* $\dim X < \infty$, $\dim Y < \infty$. *x^0 is (f, Ω, M)-extremal if and only if there exists an element $y^* \in Y^*$, $\|y^*\|_* = 1$, such that*

$$y^* \in K(f(x^0) \mid M), \tag{6.3}$$

$$0 \in D^-\langle y^*, f\rangle(x^0). \tag{6.4}$$

(iii) *Let* $Y = Y_1 \times Y_2$, $Y_1 \in \mathfrak{Y}$, $\dim Y_2 < \infty$, $M = M_1 \times M_2 \subset Y_1 \times Y_2$, $f = (f_1, f_2)$: $\Omega \to Y_1 \times Y_2$ *and a set M_1 be epi-Lipschitzian at $f_1(x^0)$. If x^0 is (f, Ω, M)-extremal then there exists an element $y^* \in Y^*$, $\|y^*\|_* = 1$, such that conditions* (6.3), (6.4) *hold.*

Theorem 6.2 gives general optimality conditions. The element y^* can be considered as an analogon of the Lagrange multipliers vector in classical nonlinear programming problem. Conditions (6.2), (6.4) generalize multipliers rule. (6.1), (6.3) correspond to conditions on a sign of multipliers and complementory slackness conditions.

Let us consider a nonlinear programming problem with equalities and inequalities: (P) $\varphi_0(x) \to \min$, $\varphi_i(x) \leqq 0$, $i = 1, 2, \ldots, m$; $\varphi_i(x) = 0$, $i = m+1, \ldots, m+n$; $x \in \Omega$. We shall suppose that Ω is closed and $f = (\varphi_0, \varphi_1, \ldots, \varphi_{m+n})$: $\Omega \to \mathbb{R}^{m+n+1}$ is Lipschitz in a neighbourhood of x^0. As a corollary of theorem 6.2 we get the following result.

Theorem 6.3: *Let x^0 be a solution of problem* (P).

(i) *For any* $\varepsilon > 0$ *there exist elements* $x \in \Omega \cap B_\varepsilon(x^0)$, $y^* \in \Lambda(x^0)$ *such that* $0 \in \partial_\varepsilon^- \langle y^*, f \rangle (x)$ *where* $\Lambda(x^0) = \left\{ y^* = (\lambda_0, \lambda_1, \ldots, \lambda_{m+n}) \in \mathbb{R}^{m+n+1} \,\middle|\, \sum_{i=0}^{m+n} |\lambda_i| = 1, \quad \lambda_i \geqq 0, \quad i = 0, 1, \ldots, m; \; \lambda_i \varphi_i(x^0) = 0, \; i = 1, 2, \ldots, m \right\}$.

(ii) *There exists an element* $y^* \in \Lambda(x^0)$ *such that* $0 \in D^- \langle y^*, f \rangle (x^0)$.

Part (ii) of theorem 6.3 is a corollary of part (i). When $\dim X < \infty$, both statements are equivalent. Theorem 6.3 remains true if all the functions are defined and LIPSCHITZ on a set Ω only. It is possible to prove necessary conditions without LIPSCHITZ assumption (cf. [26]). Using theorem 4.2 we can deduce from theorem 6.3 necessary conditions in terms of generalized differentials of individual functions.

Corollary 6.3.1: *Let* x^0 *be a solution of problem* (P), $X \in \mathfrak{Y}$.

(i) *For any* $\varepsilon > 0$ *there exist elements* $x_i \in \Omega \cap B_\varepsilon(x^0)$, $i = 0, 1, \ldots, m+n$; $\omega \in \Omega \cap B_\varepsilon(x^0)$; $y^* \in \tilde{\Lambda}(x^0)$ *such that*

$$0 \in \sum_{i=0}^{m} \lambda_i \partial_\varepsilon^- \varphi_i(x_i) + \sum_{i=m+1}^{m+n} \lambda_i \hat{\partial}_\varepsilon \varphi_i(x_i) + N_\varepsilon(\omega \mid \Omega)$$

where $\tilde{\Lambda}(x^0) = \left\{ y^* = (\lambda_0, \lambda_1, \ldots, \lambda_{m+n}) \in \mathbb{R}^{m+n+1} \,\middle|\, \sum_{i=0}^{m+n} \lambda_i = 1; \; \lambda_i \geqq 0, \; i = 0, 1, \ldots, m+n; \; \lambda_i \varphi_i(x^0) = 0, \; i = 1, 2, \ldots, m \right\}$.

(ii) *There exists an element* $y^* \in \tilde{\Lambda}(x^0)$ *such that*

$$0 \in \sum_{i=0}^{m} \lambda_i D^- \varphi_i(x^0) + \sum_{i=m+1}^{m+n} \lambda_i \hat{D} \varphi_i(x^0) + K(x^0 \mid \Omega) .$$

Unlike the classical LAGRANGE multipliers rule and its known generalizations different derivative-like objects correspond to equalities and inequalities (and the criteria) in the statement of corollary 6.3.1. Besides that all the multipliers are required to be nonnegative. The reason for the phenomenon lies in the "two-sidedness" of the equality-type constraints and the peculiarity of the definitions of $\hat{\partial}_\varepsilon \varphi(x)$ and $\hat{D}\varphi(x^0)$. Replacing them by $\partial_\varepsilon^0 \varphi(x)$ and $D^0 \varphi(x^0)$ (and thus slightly weakening the result) we can restate the multipliers rule in the more familiar way.

Corollary 6.3.2: *Let* x^0 *be a solution of problem* (P), $X \in \mathfrak{Y}$.

(i) *For any* $\varepsilon > 0$ *there exist elements* $x_i \in \Omega \cap B_\varepsilon(x^0)$, $i = 0, 1, \ldots, m+n$; $\omega \in \Omega \cap B_\varepsilon(x^0)$; $y^* \in \Lambda(x^0)$ *such that*

$$0 \in \sum_{i=0}^{m+n} \lambda_i \partial_\varepsilon^0 \varphi_i(x_i) + N_\varepsilon(\omega \mid \Omega) .$$

(ii) *There exists an element* $y^* \in \Lambda(x^0)$ *such that*

$$0 \in \sum_{i=0}^{m+n} \lambda_i D^0 \varphi_i(x^0) + K(x^0 \mid \Omega) .$$

Theorem 6.3 and its corollaries generalize and strengthen the results of

F. CLARKE, J. WARGA, J.-B. HIRIART-URRUTY and other authors [5, 11, 17, 26, 28, 41, 50]. The statements can be extended on vector and minimax optimization problems.

Example 6.1: Let us take a problem of minimizing a function $\varphi_0(x) = |x_1| - |x_2|$ under the restriction $\varphi_1(x) = 2x_1 + |x_2| = 0$, $x = (x_1, x_2) \in \mathbb{R}^2$. It is not difficult to calculate that $D^-\varphi_0(0) = \{(\alpha, \pm 1) \mid -1 \leqq \alpha \leqq 1\}$, $\hat{D}\varphi_1(0) = \{(2, \alpha) \mid -1 \leqq \alpha \leqq 1\} \cup$ $\cup (-2, -1) \cup (-2, 1)$. There could not be found $\lambda_0 \geqq 0$, $\lambda_1 \geqq 0$ at least one of them not equal to zero such that $0 \in \lambda_0 D^-\varphi_0(0) + \lambda_1 \hat{D}\varphi_1(0)$. It follows from the corollary 6.3.1 that a point 0 does not solve a problem under consideration. This conclusion could not be obtained from the theorems of F. H. CLARKE [5] (see also [17]) or J. WARGA [50].

Some examples of calculating generalized differentials are given in [26 33].

References

[1] BOLTYANSKII, V. G.: The method of tents in the theory of extremal problems. Russian Math. Surveys, **30** (1975) 3, 1–54.

[2] BOURBAKI, N.: Variétiés différentielles et analytiques. Hermann, Paris 197_.

[3] CLARKE, F. H.: Generalized gradients and applications. Trans. Amer. Math. Soc., **204** (1975) 247–262.

[4] CLARKE, F. H.: On the inverse function theorem. Pacif. J. Math., **64** (1976) 1, 97–102.

[5] CLARKE, F. H.: A new approach to LAGRANGE multipliers. Math. Oper. Res , **1** (1976) 2, 165–174.

[6] CLARKE, F. H.: Generalized gradients of LIPSCHITZ functions. Adv. Math. **40** (1981) 1, 52–67.

[7] CLARKE, F. H.: Optimization and Nonsmooth Analysis. John Wiley, New York 1983.

[8] DEMYANOV, V. F.; A. M. RUBINOV: On quasidifferentiable functionals. Sov. Math. Dokl., **12** (1980) 1, 14–17.

[9] DEMYANOV, V. F.; A. M. RUBINOV: On quasidifferentiable mappings. Math. Operationsforsch. u. Statist., Ser. Optimization, **14** (1983) 1, 3–21.

[10] DIESTEL, J.: Geometry of BANACH Spaces. Selected Topics. Springer-Verlag, Berlin e. a. 1975.

[11] DMITRUK, A. V.; A. A. MILYUTIN; N. P. OSMOLOVSKII: LYUSTERNIK's theorem and the theory of the extremum. Uspekhi Mat. Nauk, **35** (1980) 6, 11–46 (in Russian).

[12] DUBOVITSKII, A. YA.; A. A. MILYUTIN: Extremum problems in the presence of restrictions. USSR Comput. Math. and Math. Phys., **5** (1965) 3, 1–80.

[13] EKELAND, I.: On the variational principle. J. Math. Anal. Appl., **47** (1974) 2, 324–353.

[14] EKELAND, I.; G. LEBOURG: Generic FRÉCHET-differentiability and perturbed optimization problems in BANACH spaces. Trans. Amer. Math. Soc., **224** (1976) 2, 193–216.

[15] GAMKRELIDZE, R. V.: First-order necessary conditions and axioms of extremal problems. Trudy Mat. Inst. Akad. Nauk SSSR, **112** (1971), 152–180.

[16] HALKIN, H.: Interior mapping theorem with set-valued derivatives. J. Anal. Math., **30** (1976) 200–207.

[17] HIRIART-URRUTY, J.-B.: Refinements of necessary optimality conditions in nondifferentiable programming. 1. Appl. Math. Optim., **5** (1979) 1, 63–82.

[18] HIRIART-URRUTY, J.-B.: Mean value theorems in nonsmooth analysis. Numer. Funct. Anal. Optim., **2** (1980) 1, 1–30.

[19] IOFFE, A. D.: Nonsmooth analysis: differential calculus of non-differentiable mappings. Trans. Amer. Math. Soc., **266** (1981) 1, 1–56.
[20] IOFFE, A. D.: Approximate subdifferentials and applications 1: the finite dimensional theory. Trans. Amer. Math. Soc., **281** (1984) 1, 389–416.
[21] IOFFE, A. D.: Calculus of Dini subdifferentials of functions and contingent coderivatives of set-valued maps. Nonlin. Anal.: Theory, Meth. Appl., **8** (1984) 5, 517–539.
[22] IOFFE, A. D.; V. M. TICHOMIROV: Theory of Extremal problems. North-Holland, Amsterdam, 1979.
[23] KRUGER, A. YA:. Subdifferentials of nonconvex functions and generalized directional derivatives. Minsk, 1977. Deponent VINITI no. 2661–77 (in Russian).
[24] KRUGER, A. YA.: ε-semidifferentials and ε-normal elements. Minsk, 1981. Deponent VINITI no. 1331–81 (in Russian).
[25] KRUGER, A. YA.: Generalized differentials of nonsmooth functions. Minsk, 1981. Deponent VINITI no. 1332–81 (in Russian).
[26] KRUGER, A. YA.: Necessary conditions for an extremum in nonsmooth optimization problems. Minsk, 1981. Deponent VINITI no. 1333–81 (in Russian).
[27] KRUGER, A. YA.: On characterizing the covering property for nonsmooth operators. In: School on operator theory in functional spaces. Minsk, 1982, 94–95 (in Russian).
[28] KRUGER, A. YA.: Generalized differentials of nonsmooth functions and necessary conditions for an extremum. Sib. Math. J., **26** (1985) 3, 370–379.
[29] KRUGER, A. YA.: Properties of generalized differentials. Sib. Math. J., **26** (1985) 6, 1822–1832.
[30] KRUGER, A. YA.; B. SH. MORDUKHOVICH: Minimization of nonsmooth functionals in optimal control problems. Engrg. Cybernetics **16** (1978) 4, 126–133.
[31] KRUGER, A. YA.; B. SH. MORDUKHOVICH: Extremal points and the EULER equation in nonsmooth optimization problems. Dokl. Akad. Nauk RSSR **24** (1980) 8, 684–687 (in Russian).
[32] KRUGER, A. YA.; B. SH. MORDUKHOVICH: Generalized normals and derivatives and necessary conditions for an extremum in nondifferentiable programming problems. i Minsk, 1981. Deponent VINITI no. 408–80 (in Russian).
[33] KRUGER, A. YA.; B. SH. MORDUKHOVICH: Generalized normals and derivatives and necessary conditions for an extremum in nondifferentiable programming problems. ii. Minsk, 1981. Deponent VINITI no. 494–80 (in Russian).
[34] LEBOURG, G.: Valeur moyenne pour gradient généralisé. Comptes Rendus Acad. Sc. Paris, Ser. A, **281** (1975) 19, 795–798.
[35] MAGARIL-ILJAEV, G. G.: The implicit function theorem for Lipschitzian mappings. Uspekhi Mat. Naluk, **33** (1978) 1, 221–222 (in Russian).
[36] MORDUKHOVICH, B. SH.: Maximum principle in the problem of time optimal response with nonsmooth constraints. J. Appl. Math. Mech., **40** (1976) 6, 960–969.
[37] MORDUKHOVICH, B. SH.: Metric approximations and necessary optimality conditions for general classes of nonsmooth extremal problems. Sov. Math. Dokl., **22** (1980) 526–530.
[38] MORDUKHOVICH, B. SH.: Nonsmooth analysis with nonconvex generalized differentials and conjugate mappings. Dokl. Akad. Nauk BSSR, **28** (1984) 11, 976–979 (in Russian).
[39] MORDUKHOVICH, B. SH.; A. YA. KRUGER: Necessary optimality conditions in the problem of terminal control with nonfunctional constraints. Dokl. Akad. Nauk BSSR, **20** (1976) 12, 1064–1067 (in Russian).
[40] PENOT, J.-P.: Calcul sous-différentiel et optimisation. J. Funct. Anal., **27** (1978) 2, 248–276.
[41] POUREIAU, B. H.: Analysis and optimization of LIPSCHITZ continuous mappings. J. Optimiz. Theory Appl., **22** (1977) 3, 311–351.

[42] PSHENICHNIY, B. N.: Convex Analysis and Extremal Problems. Nauka, Moscow 1975 (in Russian).
[43] ROCKAFELLAR, R. T.: Convex Analysis. Princeton Univ. Press, Princeton, New Jercey 1970.
[44] ROCKAFELLAR, R. T.: Directionally Lipschitzian functions and subdifferential calculus. Proc. London Math. Soc., **39** (1979) 2, 331–355.
[45] ROCKAFELLAR, R. T.: Generalized directional derivatives and subgradients of nonconvex functions. Canad. J. Math., **32** (1980) 2, 257–280.
[46] ROCKAFELLAR, R. T.: The Theory of Subgradients and its Application to Problems of Optimization. Convex and Nonconvex Functions. Holdermann Verlag, Berlin 1981.
[47] SWEETSER, T. H.: A minimal set-valued strong derivative for vector-valued LIPSCHITZ functions. J. Optimiz. Theory Appl., **23** (1977) 4, 549–562.
[48] WARGA, J.: Derivate containers, inversefunctions and controllability. In: Calculus of Variations and Control Theory. Academic Press, New York 1976, 13–46.
[49] WARGA, J.: An implicit function theorem without differentiability. Proc. Amer. Math. Soc., **69** (1978) 1, 65–69.
[50] WARGA, J.: Controllability and a multiplier rule for nondifferentiable optimization problems. SIAM J. Control Optimiz., **16** (1978) 5, 803–812.
[51] WARGA, J.: Fat homeomorphisms and unbounded derivate containers. J. Math. Anal. Appl., **81** (1981) 545–560.
[52] WARGA, J.: On bounding, interior and covering functions. J. Math. Anal. Appl., **82** (1981) 255–267.

Received June 1986, revised May 1987

A. Ya. Kruger
PTO "Avtotranssystema"
Chapaeva 5
220600 Minsk
USSR

optimization 19 (1988) 6, 781–802

A Continuous Set Covering Problem as a Quasidifferentiable Optimization Problem

H. JANDL and K. WIEDER

Universität Karlsruhe, Institut für Statistik und Mathematische Wirtschaftstheorie

Summary: In this paper we present an algorithm to solve a family of finite covering problems in $\mathbb{R}^2$.

Given a compact, finitely convex decomposable set $\Omega \subset \mathbb{R}^2$ and an integer $m \in \mathbb{N}$, we are looking for the centers and the minimal radius of m balls $B(x_1, r), \ldots, B(x_m, r)$ with the property $\Omega \subset \bigcup_{i=1}^{m} B(x_i, r)$, $x_i \in \Omega$.

It will be shown that this problem can be reduced to the computation of DIRICHLET tessellations (VORONOI sets) and the computation of minima of quasidifferentiable optimization problems.

AMS 1980 Subject Classification:

Primary: 90 C 30; Secondary: 26 B 05.

Key words: Set covering, VORONOI sets, DIRICHLET tesselation, quasidifferentiable functions, locally LIPSCHITZ functions.

0. Introduction

Considerable practical interest in *nonsmooth optimization* has been generated by a variety of publications. The main purpose of these publications was to develop algorithmic approaches to solve restricted and unrestricted nonsmooth optimization problems. These problems arise if the objective function or the set of feasible solutions is given by not necessarily continuously differentiable real-valued functions.

A promising approach was proposed by V. F. DEMYANOV and A. M. RUBINOV, in 1983. In their paper they considered the class of *quasi-differentiable functions* where the (quasi-) differential (the analogy of the classical gradient) can be represented by a pair of compact, convex sets of $\mathbb{R}^n$. This pair of sets (the *quasi-differential*) predicates descent or ascent directions, which give way to gradient type methods as already used for smooth optimization problems.

An important property of the class of quasi-differentiable functions is its closedness with respect to all algebraic operations and additionally to taking pointwise maxima and minima.

In this paper an application to the theory of quasi-differentiable optimization is considered. More precisely, it will be shown how this calculus may be used to obtain minimal coverings of compact polyhedral sets in $\mathbb{R}^2$, which are convexly decomposable.

The paper is organized as follows:

In Chapter 1 the definition and a short overview of important properties of quasi-differentiable functions are given. A way to compute local minima of restricted and unrestricted quasi-differentiable optimization problems is presented in Chapter 2.

The reminder of this paper deals with the description, the properties of a continuous set covering problem and its decomposition into quasi-differentiable optimization problems. An example is the contents of the last chapter.

1. Calculus Quasidifferentialis

Throughout this paper we will use the following notations:

$\|.\|$ stands for the usual Euclidean norm. By $\Phi|_x$ we denote the value of the function Φ taken at x. $P(A)$ is the set of all subsets of A and $B(x_0, r) := := \{x \in \mathbb{R}^n \mid \|x_0 - x\| \leqq r\}$.

The calculus of quasi-differentiable functions is described to its full extent in [2]. In the following a brief description of this calculus is given, as far as it will be needed for subsequent statements and arguments.

Definition 1.0: A real-valued function Φ defined on an open set $S \subseteqq \mathbb{R}^n$ is said to be *quasi-differentiable* at a given point $x_0 \in S$, if it is directionally differentiable at x_0 and if there exist convex and compact sets $\underline{\partial}\Phi|_{x_0} \subset \mathbb{R}^n$ and $\overline{\partial}\Phi|_{x_0} \subset \mathbb{R}^n$ such that for every $g \in \mathbb{R}^n$

$$\left.\frac{\partial \Phi}{\partial g}\right|_{x_0} := \lim_{\alpha \downarrow 0} \frac{1}{\alpha} \cdot [\Phi(x_0 + \alpha \cdot g) - \Phi(x_0)] \tag{1.0}$$

$$= \sup_{v \in \underline{\partial}\Phi|_{x_0}} \langle v, g \rangle + \inf_{w \in \overline{\partial}\Phi|_{x_0}} | \langle w, g \rangle .$$

The pair $D\Phi|_{x_0} := [\underline{\partial}\Phi|_{x_0}, \overline{\partial}\Phi|_{x_0}]$ is called the *quasi-differential*, $\underline{\partial}\Phi|_{x_0}$ and $\overline{\partial}\Phi|_{x_0}$ are called respectively, the *sub-differential* and the *super-differential* of Φ at x_0.

Remark 1.0: The quasi-differential is not unique. If A is a convex compact subset of $\mathbb{R}^n$ then $[\underline{\partial}\Phi|_{x_0}, \overline{\partial}\Phi|_{x_0}]$ and $[\underline{\partial}\Phi|_{x_0} + A, \overline{\partial}\Phi|_{x_0} - A]$ are both quasi-differentials satisfying (1.0).

Definition 1.1: A real-valued function Φ defined on an open set $S \subseteqq \mathbb{R}^n$ is said to be *uniformly quasi-differentiable* at a given point $x_0 \in S$, if it is quasi-differentiable and uniformly directionally differentiable at x_0.

The following theorem does not only state that the class of quasi-differentiable functions is a linear space and closed with respect to taking pointwise maxima and minima, but also specifies the quasi-differentials obtained by these operations. It is therefore of major significance to the theory of quasi-differentiable functions.

Theorem 1.0: *Let Φ_i, $i \in I := \{1, \ldots, m\}$ be quasi-differentiable at $x_0 \in S \subseteqq \mathbb{R}^n$ and let $c \in \mathbb{R}$, then the following statements hold:*

1. $\Phi := c \cdot \Phi_1$ *is quasi-differentiable and* $D\Phi|_{x_0}$ *is given by*
$D\Phi|_{x_0} := [c \cdot \underline{\partial}\Phi_1|_{x_0}, c \cdot \bar{\partial}\Phi_1|_{x_0}]$ *if* $c \geqq 0$,
$D\Phi|_{x_0} := [c \cdot \bar{\partial}\Phi_1|_{x_0}, c \cdot \underline{\partial}\Phi_1|_{x_0}]$ *if* $c < 0$.
2. $\Phi := \Phi_1 + \Phi_2$ *is quasi differentiable and* $D\Phi|_{x_0}$ *is given by*
$D\Phi \mid x_0 := [\underline{\partial}\Phi_1|_{x_0} + \partial\Phi_2|_{x_0}, \bar{\partial}\Phi_1|_{x_0} + \bar{\partial}\Phi_2|_{x_0}]$.
3. $\Phi := \inf_{i \in I} \Phi_i$ *is quasi-differentiable and* $D\Phi|_{x_0}$ *is given by*
$D\Phi|_{x_0} := [\partial\Phi|_{x_0}, \bar{\partial}\Phi|_{x_0}]$, *where*
$\underline{\partial}\Phi|_{x_0} := \sum_{k \in R|_{x_0}} \underline{\partial}\Phi_k|_{x_0}$, *and*
$\bar{\partial}\Phi|_{x_0} := \operatorname{conv}\,[\bar{\partial}\Phi_k|_{x_0} + \underline{\partial}\Phi_k|_{x_0} - \sum_{\substack{i \in R|_{x_0} \\ i \neq k}} \underline{\partial}\Phi_i|_{x_0} : k \in R|_{x_0}]$, *where*
$R: S \to P(I)$, $R(x) := \{i \in I : \Phi|_{x_0} = \Phi_i|_{x_0}\}$.

Proof: The proofs are straightforward. They follow from Definition 1.0. ∎

Additional properties, as the closedness with respect to taking the pointwise supremum of a finite number of quasi-differentiable functions can be derived easily from Theorem 1.0.

Lemma 1.0: *Let Φ be quasi-differentiable at $x_0 \in S \subseteqq \mathbb{R}^n$, S open. A steepest descent direction g^* of Φ at x_0 is given by*

$$g^*|_{x_0} := -\frac{w_0 - v_0}{\|w_0 - v_0\|},$$

where

$$\|w_0 - v_0\| := \sup_{w \in \bar{\partial}\Phi|_{x_0}} \inf_{v \in -\underline{\partial}\Phi|_{x_0}} \|w - v\|. \tag{1.1}$$

Proof: The proof is straightforward and is omitted here. ∎

The vector $-(w_0 - v_0)$ is the analogy of the classical steepest descent direction. In constrast to continuously differentiable functions however, the direction of steepest descent of a quasi-differentiable function is not necessarily unique.

Lemma 1.1: *Let Φ be quasi-differentiable at $x_0 \in S \subseteqq \mathbb{R}^n$. Using the notations of Lemma 1.0, there exists an extremal point $w_0 \in \bar{\partial}\Phi|_{x_0}$ and a $v_0 \in -\underline{\partial}\Phi|_{x_0}$ satisfying (1.1).*

Proof: The proof uses standard arguments of convex optimization and is omitted here. ∎

Lemma 1.2: *Let Φ be quasi-differentiable at $x_0 \in S \subseteqq \mathbb{R}^n$. For x_0 to be the argument of a local minimum of Φ it is neccessary that*

$$-\bar{\partial}\Phi\,|_{x_0} \subset \underline{\partial}\Phi\,|_{x_0}\,. \tag{1.3}$$

Remark 1.1: Points $x_0 \in S$ satisfying (1.3) are said to be *inf-stationary.*

A sufficient condition for x_0 to be the argument of a strict local minimum of a uniformly quasi-differentiable function Φ is given by

Lemma 1.3: *Let Φ be uniformly quasi-differentiable at $x_0 \in S \subseteqq \mathbb{R}^n$. For x_0 to be the argument of a strict local minimum of Φ it is sufficient that*

$$-\bar{\partial}\Phi\,|_{x_0} \subset \operatorname{int}\,(\underline{\partial}\Phi\,|_{x_0})\,. \tag{1.4}$$

Proofs for Lemmas 1.2 and 1.3 are given in Theorem 16.2 and Theorem 16.7 in [2].

Note that if (1.4) holds, we already obtain a sufficient condition derived from an extension of a first order derivative, which is impossible if $\Phi \in C^1(B(x_0, \varepsilon))$.

2. Computation of Local Minima of Quasi-Differentiable Functions

For the following let us assume that M is a set which can be described by the finite intersection of level sets of quasi-differentiable functions: $h_i: \mathbb{R}^n \to \mathbb{R}$, $i \in I := \{1, \ldots, m\}$, i.e.

$$M := \{x \in \mathbb{R}^n \mid h(x) \leqq 0\}\,, \quad \text{where}$$
$$h:\ \mathbb{R}^n \to \mathbb{R}$$
$$x \mapsto \sup_{i \in I} h_i(x)\,.$$

Furthermore let us assume that a function $\Phi: \mathbb{R}^n \to \mathbb{R}$ is quasi-differentiable on int (M) and that its quasi-differential is given by the convex hulls of finite and nonempty sets of points, i.e.

$$\bar{\partial}\Phi\,|_{x_0} := \operatorname{conv}\,\{w_1, \ldots, w_r\} \quad \text{and}$$
$$-\underline{\partial}\Phi\,|_{x_0} := \operatorname{conv}\,\{v_1, \ldots, v_s\}\,, \quad r, s \in \mathbb{N}\,.$$

Local minima can be found by a so-called descent method. This is to generate a sequence $\{x_k\}_{k \in \mathrm{N}}$ by defining

$$x_{k+1} := x_k + \alpha_k \cdot g^*\,|_{x_k}\,,$$

where $g^*\,|_{x_k} \in \mathbb{R}^n$ is a steepest descent direction of Φ at $x_k \in$ int (M) and α_k may be any "suitable" stepsize that guarantees $x_{k+1} \in M$ and

$$\Phi\,(x_{k+1}) < \Phi(x_k)\,.$$

The selection of the stepsize α_k is similar to that in smooth optimization. However, and this is obvious, one must not use a method that assumes the differ ent ability of Φ. In the sequel we are only concerned with the determination of a steepest descent direction $g^* |_{x_k}$.

If x_k is an element of the boundary of $M\,(bd(M))$ we face the usual problem of finding a *feasible* descent direction of Φ at x_k. Assume that M satisfies a regularity condition at x_k: $\forall \varepsilon > 0\ \exists x \in B(x_k, \varepsilon)$ with $h(x) < 0$. As proposed in [3] the objective function may then be locally (at x_k) replaced by Φ_{x_k}, where

$$\Phi_{x_k} : \mathbb{R}^n \to \mathbb{R}$$
$$x \mapsto \sup\,\{\Phi(x) - \Phi\,|_{x_k}\,,\, h(x)\}\,.$$

By Theorem 1.0 Φ_{x_k} is quasi-differentiable and a steepest descent direction of Φ_{x_k} at x_k is given by Lemma 1.0. It is clear that such a descent direction is not a steepest with respect to Φ but at least it is a feasible descent direction. This is the analogy of the method of reduced gradients as used in smooth optimization.

In the given assumptions the sets of extremal points of the super-differentials of Φ and Φ_{x_k} (ext(.)), are finite for every $x_k \in \text{int}(M)$ and for every $x_k \in \text{bd}(M)$. Due to Lemma 1.1, the computation of steepest descent directions of Φ or Φ_{x_k} at x_k is a problem of finding orthogonal projections of the origin into compact polyhedrons given by the convex hulls of finite numbers of points, since (1.1) can be written as

$$\|w_0 - v_0\| := \sup_{w \in \text{ext}(\bar{\partial}\Phi|_{x_k})} \Big(\inf_{v \in -\{\underline{\partial}\Phi|_{x_0} + \{w\}\}} \|v\| \Big)$$

or if $x_k \in \text{bd}(M)$

$$\|w_0 - v_0\| := \sup_{w \in \text{ext}(\bar{\partial}\Phi_{x_k}|_{x_k})} \Big(\inf_{v \in -\{\underline{\partial}\Phi_{x_k}|_{x_k} + \{w\}\}} \|v\| \Big)\,.$$

For simplification let $V := -\ \{v_1 + w, \ldots, v_s + w\}$, $w \in \text{ext}\,(\bar{\partial}\Phi\,|_{x_0})$ and consider the function

$$\psi : \text{conv}\,(V) \to \mathbb{R}$$
$$v \mapsto \psi(v) := \|v\|^2\,.$$

Obviously we have a convex optimization problem. The difficulty however is due to the representation of the set of feasible solutions. It is not given by the finite intersection of affine halfspaces. Changing the representation into something like

$$\text{conv}\,(V) := \{x \in \mathbb{R}^n \mid \langle a_j, x\rangle \leqq b_j\,,\ a_j \in \mathbb{R}^n\,,\ b_j \in \mathbb{R}\,,\ j = 1, \ldots, t,\ t \in \mathbb{N}\}\,,$$

would make it possible to use standard algorithms known from convex optimization. In general, however this proceeding is not advantageous without the knowledge of additional information. It should be mentioned in this context that usually $\{v_1, \ldots, v_s\} \neq \text{ext}\,(\text{conv}\,\{v_1, \ldots, v_s\})$. Therefore we suggest the following algorithm to minimize ψ, which is a modification of an algorithm suggested by V. DEMYANOV.

Algorithm I

Initialization: Let $j=0$, $V_0:=V$, $\mu_0\in\text{conv}(V)$, and let $\varepsilon>0$ be sufficiently small.

Step 1: If $\|\mu_j\|^2<\varepsilon^2$ then $0\in\text{conv}(V)+\{x\in\mathbb{R}^n \mid \|x\|=\varepsilon\}$. In this case let $\mu_{j+1}:=\mu_j$ and continue at Step 4.
Determine

$$H_j:=\{h\in V_j \mid \langle\mu_j, \mu_j-h\rangle>0\}.$$

If $H_j=\emptyset$ then let $\mu_{j+1}:=\mu_j$ and continue at Step 4.

Step 2: Compute

$$(t_j^*, h_j^*):=\operatorname*{arginf}_{(t,h)\in T_j}[t^2\|\mu_j-h\|^2-2\cdot t\langle\mu_j, \mu_j-h\rangle+\|\mu_j\|^2], \quad \text{where}$$

$$T_j:=\left\{(t, h)\in\mathbb{R}\times H_j \mid t=\inf\left(1, \frac{\langle\mu_j, \mu_j-h\rangle}{\|\mu_j-h\|^2}\right)\right\}. \tag{2.0}$$

Step 3: Let $\mu_{j+1}:=t_j^*\cdot h_j^*+(1-t_j^*)\cdot\mu_j$.
If $\|\mu_{j+1}-\mu_j\|^2<\varepsilon^2$ then continue at Step 4.
$V_{j+1}:=V_0\cup\{\mu_{j+1}\}$.
$j=j+1$.
Continue at Step 1.

Step 4: μ_{j+1} minimizes ψ.

Remark to Step 1: H_j and $\{0\}$ are subsets of the same open affine halfspace

$$\{x\in\mathbb{R}^n \mid \langle\mu_j, \mu_j-x\rangle>0\}.$$

Remark to Step 2: Choose $(t_j^*, h_j^*)\in\mathbb{R}\times H_j$ in a way that the distance of the origin to the line segments $[\mu_j, h]$, $h\in H_j$ is minimized.
This is achieved by minimizing the function

$$\begin{aligned} g_j&:[0, 1]\times H_j\to\mathbb{R}\\ &(t, h)\mapsto g_j(t, h):=\|t\cdot h+(1-t)\cdot\mu_j\|^2. \end{aligned} \tag{2.1}$$

A straightforward computation shows that (2.1) can be written as

$$g_j(t, h)=t^2\|\mu_j-h\|^2-2\cdot t\langle\mu_j, \mu_j-h\rangle+\|\mu_j\|^2.$$

Equating $\frac{dg_j(t, h)}{dt}$ with zero and keeping in mind that $t\in[0, 1]$ we obtain

$$t:=\inf\left(1, \sup\left(\frac{\langle\mu_j, \mu_j-h\rangle}{\|\mu_j-h\|^2}\right)\right). \tag{2.2}$$

From the definition of H_j it is clear that $\langle\mu_j, \mu_j-h\rangle>0$ and therefore (2.2) reduces to

$$t:=\inf\left(1, \frac{\langle\mu_j, \mu_j-h\rangle}{\|\mu_j-h\|^2}\right).$$

Lemma 2.0: *The sequence* $\{\mu_k\}_{k\in\mathbb{N}}$ *generated by Algorithm I converges and*

$$\lim_{k\to\infty} \mu_k = v_0 .$$

Proof: The sequence is bounded from below by the optimal solution v_0. We show that it is strictly monotonously decreasing, i.e.

$$\|\mu_{j+1}\|^2 = t^2 \cdot \|\mu_j - h\|^2 - 2t \cdot \langle \mu_j, \mu_j - h\rangle \|\mu_j\|^2 < \|\mu_j\|^2 . \tag{2.3}$$

Since $t>0$, relation (2.3) is equivalent to

$$t \cdot \|\mu_j - h\|^2 < 2 \cdot \langle \mu_j, \mu_j - h\rangle .$$

This is obviously correct for every $h \in H_j$, since

$$0 < t = \inf\left(1, \frac{\langle \mu_j, \mu_j - h\rangle}{\|\mu_j - h\|^2}\right) \leqq \frac{\langle \mu_j, \mu_j - h\rangle}{\|\mu_j - h\|^2} < 2 \cdot \frac{\langle \mu_j, \mu_j - h\rangle}{\|\mu_j = h\|^2} .$$

We will lead the assertation that

$$\lim_{k\to\infty} \mu_k \neq v_0$$

to a contradiction.
Take an arbitrary $k \in \mathbb{N}$, $\mu_k \neq v_0$ and assume that H_k is empty. Hence

$$\langle \mu_k, \mu_k - v_i\rangle \leqq 0 , \quad \forall v_i \in V . \tag{2.4}$$

Furthermore $\|v_0\|^2 < \|\mu_k\|^2$ implies

$$\langle \mu_k, \mu_k - v_0\rangle > 0$$

and since $v_0 \in \operatorname{conv}(V)$

$$\left\langle \mu_k, \mu_k - \sum_{i=1}^{s} \lambda_i v_i \right\rangle = \sum_{i=1}^{s} \lambda_i \langle \mu_k, \mu_k - v_i\rangle > 0 ,$$

which contradicts (2.4) and completes the proof. ∎

Remark 2.0: Together with Lemma 2.0, the stopping rule in Step 3 of Algorithm I guarantees that the algorithm stops after a finite number of iterations.

Remark 2.1: For further information see also [6] and [7].

3. The Finite Ball Covering of Compact Sets

Definition 3.0: A system of sets A is called a *Covering* of a set $\Omega \subset \mathbb{R}^n$ if for al $x \in \Omega$ there exists a set $C \in A$ with $x \in C$.

Definition 3.1: Let $I := \{1, \ldots, m\}$ be a nonempty and finite set of positive integers. A covering $A(x, r)$ of a set $\Omega \subset \mathbb{R}^n$ is said to be a *Finite Ball Covering* generated by $x := (x_1, \ldots, x_m)$ with radius r if

$$A(x, r) := \{B(x_i, r\} \mid x_i \in \mathbb{R}^n , \ r \in \mathbb{R}^+ , \ i \in I\} ,$$

and it is called a *restricted Finite Ball Covering* generated by x with radius r if

$$A(x, r) := \{B(x_i, r) \mid x_i \in \Omega, \ r \in \mathbb{R}^+ , \ i \in I\} .$$

Three interesting problems are raised by Definition 3.1.

1. Find a "minimal ball covering $A(x, r)$ of Ω generated by x", i.e. find

$$\inf \{r \in \mathbb{R}^+ \mid A(x, r) \text{ is a covering of } \Omega\},$$

2. find a "minimal ball covering $A(x, r)$ of Ω", i.e. find

$$\inf \{r \in \mathbb{R}^+ \mid x \in (\mathbb{R}^n)^m \text{ and } A(x, r) \text{ is a covering of } \Omega\}, \text{ and}$$

3. find a "minimal restricted ball covering $A(x, r)$ of Ω", i.e. find

$$\inf \{r \in \mathbb{R}^+ \mid x \in \Omega^m \text{ and } A(x, r) \text{ is a covering of } \Omega\}.$$

If Ω is bounded, the minimization of the radius r is of no interest since I is finite. Therefore we restrict ourselves to the covering of compact sets.

Lemma 3.0: *Given a compact set $\Omega \subseteqq \mathbb{R}^n$, a finite set of positive integers $I := \{1, \ldots, m\}$ and a fixed vector $x := (x_1, \ldots, x_m) \in (\mathbb{R}^n)^m$ then*

$$A(x, r_0) := A(x, \sup_{\xi \in \Omega} \inf_{i \in I} \|x_i - \xi\|)$$

is a minimal ball covering of Ω generated by x.

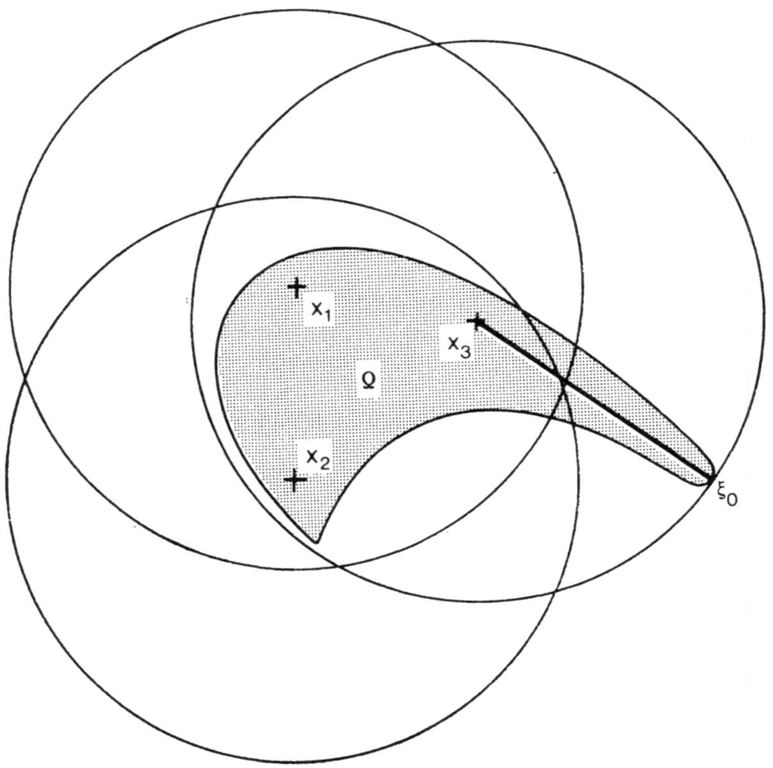

Fig. 3.0: A minimal covering of Ω generated by $x = (x_1; x_2, x_3)$. $\Gamma(x) = \|x_3 - \xi_0\|$, $A(x, r) = \{B(x_i, \|x_3 - \xi_0\|) \mid i \in I\}$.

Proof: It is obvious that $A(x, r_0)$ is a covering of Ω. We show that it is minimal with respect to x. Since Ω is compact and I is finite there exists a set $K \subseteqq I$ such that $\sup_{\xi\in\Omega} \inf_{i\in I} \|x_i - \xi\| = \|x_k - \xi_0\| = \|x_k - \xi_0\| \, \forall k \in K$ and $\xi_0 \in \mathrm{bd}(B(x_k, \|x_k - \xi_0\|))$. Let $r^* < \|x_k - \xi_0\|$, $k \in K$. Then for every $k \in K$ we can strictly separate ξ_0 and $B(x_k, r^*)$ and since for every $i \in I \backslash K$ $\|x_i - \xi_0\| \geqq \|x_k - \xi_0\| > r^*$, ξ_0 is also strictly separable from $B(x_i, r^*)$, which implies that $\xi_0 \notin A(x, r^*)$ and completes the proof. ■

We will now formulate Problem 2 and 3 as a minimization problem. Let Ω be a compact subset of $\mathbb{R}^n$ and let $I := \{1, \ldots, m\}$ be a nonempty and finite set of positive integers, then Problem 2 (3) is equivalent to:

Minimize

$$\Gamma : (\mathbb{R}^n)^m \to \mathbb{R}$$
$$(x_1, \ldots, x_m) \mapsto \Gamma(x) := \sup_{\xi\in\Omega} \inf_{i\in I} \|x_i - \xi\| \quad (\text{s.t. } x \in \Omega^m) .$$

We will call this minimization problem a *finite ball covering problem.*

4. Properties of the Finite Ball Covering Problem

In this chapter we will discuss the properties of the restricted finite ball covering problem. First we state that it is Lipschitzian and directionally differentiable. In addition we present methods to compute function values and descents with the help of the calculus quasidifferentialis.

Theorem 4.0: *Given a compact set $\Omega \subset \mathbb{R}^n$ and a finite set of positive integers $I := \{1, \ldots, m\}$, I, $\Omega \neq \emptyset$, the function*

$$\Gamma : (\mathbb{R}^n)^m : \to \mathbb{R}$$
$$(x_1, \ldots, x_m) \mapsto \Gamma(x) := \sup_{\xi\in\Omega} \inf_{i\in I} \|x_i - \xi\| ,$$

is Lipschitzian on Ω.

Proof: Since Ω is assumed to be compact it sufficies to prove that Ω is locally Lipschitzian for every x of its domain.

Consider the function

$$\Phi : (\mathbb{R}^n)^m \times \mathbb{R}^n \to \mathbb{R}$$
$$(x_1, \ldots, x_m, \xi) \mapsto \Phi(x, \xi) := \inf_{i\in I} \|x_i - \xi\| .$$

We have to show that for all $(x, \lambda) \in \mathrm{dom}(\Phi)$ and for all sequences $\{(y_j, \mu_j)\}_{j\in\mathbb{N}} \xrightarrow[j\to\infty]{} (x, \lambda)$ the following relation holds:

$$\lim_{j\to\infty} \frac{|\Phi(y_j, \mu_j) - \Phi(x, \lambda)|}{\|(y_j, \mu_j) - (x, \lambda)\|} \leqq L , \tag{4.0}$$

where $(y_j, \mu_j) \in \mathrm{dom}(\Phi)$.

For the numerator we have

$$|\Phi(y_j, \mu_j) - \Phi(x, \lambda)| = |\inf_{i\in I} \|y_{ji} - \mu_j\| - \inf_{i\in I} \|x_i - \lambda\||. \tag{4.1}$$

Since I is a finite set there exists a set $K \subseteqq I$ such that for all $k \in K$ $\|x_k - \lambda\| = \inf_{i\in I} \|x_i - \lambda\|$. For every sequence $\{(y_j, \mu_j)\}_{j\in \mathbb{N}} \xrightarrow[j\to\infty]{} (x, \lambda)$, we know by definition that for every $\varepsilon > 0$ there exists a $j(\varepsilon) \in \mathbb{N}$ such that for every $j > j(\varepsilon)$ and for all $i \in I$, $\|y_{ji} - x_i\| \leqq \varepsilon$ and $\|\mu_j - x\| \leqq \varepsilon$. Now choose $\varepsilon > 0$ in a way such that for an arbitrary $k \in K$

$$B(\lambda, \|x_k - \lambda\| + \varepsilon) \cap \bigcup_{i\in I\setminus K} B(x_i, \varepsilon) = \varnothing$$

If $I = K$ then $\bigcup_{i\in I\setminus K} B(x_i, \varepsilon) := \varnothing$ and let $j(\varepsilon) = 1$.

Then for every $j > j(\varepsilon)$ we may rewrite (4.1) as

$$\begin{aligned} \inf_{i\in K} \|y_{ji} - \mu_j\| - \inf_{i\in K} \|x_i - \lambda\|| &= |\inf_{i\in K} [\|y_{ji} - \mu_j\| - \|x_i - \lambda\|]| \\ &\leqq \sup_{i\in K} |\|y_{ji} - \mu_j\| - \|x_i - \lambda\||. \end{aligned}$$

Since K is a finite set there exists a $k \in K$ such that

$$\begin{aligned} \sup_{i\in K} |\|y_{ji} - \mu_j\| - \|x_i - \lambda\|| &= |\|y_{jk} - \mu_j\| - \|x_k - \lambda\|| \\ &\leqq \|(y_{jk} - x_k) - (\mu_j - \lambda)\|. \end{aligned}$$

For the denominator we have

$$\|(y_j, \mu_j) - (x, \lambda)\| = \|((y_j - x), (\mu_j - \lambda))\| \geqq \|(y_{jk} - x_k), (\mu_j - \lambda)\|.$$

For simplification let $a := y_{jk} - x_k$ and $b := \mu_j - \lambda$. Thus is remains to show, that $\|a - b\| \leqq \|(a, b)\|$ or $\|a\|^2 - 2 \cdot \|a\| \cdot \|b\| + \|b\|^2 \leqq (\|a\|^2 + \|b\|^2)$, which holds true for every $L \geqq 1$.

Thus relation (4.0) holds true for every $(x, \lambda) \in \operatorname{dom}(\Phi)$, especially for

$$((x_1, \ldots, x_m), \operatorname*{argsup}_{\xi\in\Omega} \inf_{i\in I} \|x_i - \xi\|) \in \operatorname{dom}(\Phi).$$

Since $\|((y, \mu) - (x, \lambda))\| \geqq \|y - x\|$ we find that also Γ is Lipschitzian, which completes the proof. ■

Theorem 4.1: *Given a compact quasi-differentiable set $\Omega \subset \mathbb{R}^n$ i.e. a set which can be described by the intersection of level sets of quasi-differentiable functions defined on $\mathbb{R}^n$, a finite set of positive integers $I := \{1, \ldots, m\}$, $I, \Omega \neq \varnothing$ and a function*

$$\begin{aligned} &\Gamma: (\mathbb{R}^n)^m \to \mathbb{R} \\ &(x_1, \ldots, x_m) \mapsto \Gamma(x) := \sup_{\xi\in\Omega} \inf_{i\in I} \|x_i - \xi), \end{aligned}$$

the rate of ascent at x in direction $g := (g_1, \ldots, g_m)$ is obtained by

$$\left.\frac{\partial \Gamma}{\partial g}\right|_{x_0} = \sup_{\xi\in S|_{x_0}} \sup_{q\in Q|_\xi} \frac{\partial \Phi(x_0, \xi)}{\partial(g, q)},$$

where

1. $\Phi:(\mathbb{R}^n)^m\times\mathbb{R}^n\to\mathbb{R}$

$$(x_1,\ldots,x_m,\xi)\mapsto\Phi(x,\xi):=\inf_{i\in I}\|x_i-\xi\|\,,$$

2. $S:(\mathbb{R}^n)^m\to P(\Omega)$

$$(x_1,\ldots,x_m)\mapsto S(x):=(\xi\in\Omega\mid\Phi|_{(x,\xi)}=\Gamma|_x\}\,,$$

3. $Q:\Omega\to P(\mathbb{R}^n))$

$$\xi\mapsto Q(\xi):=\{g\in\mathbb{R}^n\mid\exists\alpha>0\text{ such that }\xi+\alpha\cdot g\in\Omega\,\forall\alpha\in[0,\alpha_0]\}\,.$$

Proof: The proof is a direct consequence of Theorem 1 in [4]. ■

The computation of a function value of a finite ball covering problem as formalized in the previous chapter is an optimization problem. In the sequel we will show that there exists a different representation of the finite ball covering problem which enables us, under additional assumptions, to compute function values of the objective function. To show that these representations are equivalent we need

Definition 4.0: Let $I:=\{1,\ldots,m\}$, $I\neq\emptyset$ be a finite set of positive integers and let D be a set-valued mapping defined by

$$D:(\mathbb{R}^n)\to P((\mathbb{R}^n)^m)$$
$$(x_1,\ldots,x_m)\mapsto D(x):=\{D_1(x),\ldots,D_m(x)\}\,,$$

where

$$D_i(x):=\{y\in\mathbb{R}^n\mid\|x_i-y\|=\inf_{j\in I}\|x_j-y\|\}\,.$$

Then $D(x)$ is called the DIRICHLET *tessellation of* $\mathbb{R}^n$ *generated by* x.

Remark 4.0: $D(x)\cap\Omega:=\{D_1(x)\cap\Omega,\ldots,D_m(x)\cap\Omega\}=\{\Omega(x_1),\ldots,\Omega(x_m)\}$ will be called the DIRICHLET *tessellation of* Ω *generated by* x. If confusion is not possible we will also use the notation $\{\Omega_1,\ldots,\Omega_m\}$.

Remark 4.1: The sets $D_i(x)$ (or Ω_i) are also known as VORONOI sets.

Lemma 4.0: *Given a compact set* $\Omega\subset\mathbb{R}^n$ *and a finite set of positive integers* $I:=\{1,\ldots,m\}$, I, $\Omega\neq\emptyset$ *and* $x_k\in\mathbb{R}^n\,\forall k\in I$, *then the following statements hold:*

1. $$\Omega=\bigcup_{i\in I}\Omega_i\,,$$

where

$$\Omega_i:=\{\xi\in\Omega\mid\|x_i-\xi\|=\inf_{j\in I}\|x_j-\xi\|\}\,.$$

2. *The set* Ω_i *is compact for every* $i\in I$.

Proof: The proofs follow from standard arguments and are omitted here. ■

Lemma 4.1: *Given a compact set* $\Omega\subset\mathbb{R}^n$ *and a finite set of positive integers* $I:=\{1,\ldots,m\}$ I, $\Omega\neq\emptyset$ and $x_k\in\Omega\ \forall k\in I$, *the following equation holds:*

$$\sup_{\xi\in\Omega}\inf_{i\in I}\|x_i-\xi\|=\sup\,[\sup_{\xi\in\Omega_i}\|x_i-\xi\|\mid i\in I]\qquad(4.2)$$

where

$$\Omega_i := \{\xi \in \Omega \mid \|x_i - \xi\| = \inf_{j \in I} \|x_j - \xi\|\} .$$

Proof: For every $\xi \in \Omega$ there exists a $k \in I$ such that $\inf_{j \in I} \|x_j - \xi\| = \|x_k - \xi\|$ since I is a finite set. This holds true for a ξ_0, where

$$\Phi(x) = \inf_{j \in I} \|x_j - \xi_0\| = \|x_k - \xi_0\| .$$

Suppose there exists a $\mu \in \Omega_k$ such that $\|x_k - \mu\| > \|x_k - \xi\|$. This would imply

$$\begin{aligned} \Phi(x) &= \inf \{\|x_1 - \xi_0\|, \ldots, \|x_k - \xi_0\|, \ldots, \|x_m - \xi_0\|\} \\ &< \inf \{\|x_1 - \mu\|, \ldots, \|x_k - \mu\|, \ldots, \|x_m - \mu\|\} \\ &= \|x_k - \mu\| , \quad \text{since } \mu \in \Omega_k . \end{aligned}$$

This however contradicts the definition of Φ. Thus

$$\|x_k - \xi_0\| = \sup_{\xi \in \Omega_k} \|x_k - \xi\| = \Phi(x) .$$

Furthermore assume that there exists an index $l \in I$ such that

$$\sup_{\xi \in \Omega} \|x_l - \xi\| = : \|x_l - \xi_l\| > \|x_k - \xi_0\| := \Phi(x) .$$

This however leads to the following contradiction:

$$\begin{aligned} \Phi(x) &= \sup_{\xi \in \Omega} \inf_{i \in I} \|x_i - \xi\| \\ &\geqq \sup_{\xi \in \{\xi_1, \xi_0\}} \inf_{i \in I} \|x_i - \xi\| \\ &= \sup \{\inf_{i \in I} \|x_i - \xi_l\|, \inf_{i \in I} \|x_i - \xi_0\|\} \\ &= \sup \{\|x_l - \xi_l\|, \|x_k - \xi_0\|\} \\ &= \|x_l - \xi_l\| > \|x_k - \xi_0\| = \Phi(x) . \end{aligned}$$

Hence $\Phi(x) = \|x_k - \xi_0\| = \sup [\sup_{\xi \in \Omega_i} \|x_i - \xi\| \mid i \in I]$, which completes the proof. ■

Remark 4.2: Keep in mind that the sets Ω_i of (4.2) depend on the choice of $x := (x_1, \ldots, x_m)$.

Lemma 4.2: *Given a nonempty compact set $A \subset \mathbb{R}^n$ then for every $x \in \mathbb{R}^n$ the following statement holds:*

$$\xi_0 := \sup_{\xi \in A} \|x - \xi\| = \sup_{\xi \in \mathrm{conv}(A)} \|x - \xi\|$$

Proof: The proof uses standard arguments from convex analysis and is omitted here. ■

We will summarize these statements in

Corollary 4.0: *Given compact set $\Omega \subset \mathbb{R}^n$, a finite set of positive integers $I := \{1, \ldots, m\}$ I, $\Omega \neq 0$ and functions Γ_1, Γ_2 defined by*

$$\Gamma_1 : (\mathbb{R}^n)^m \to \mathbb{R}$$

$$(x_1, \ldots, x_m) \mapsto \Gamma_1(x) := \sup_{\xi \in \Omega} \inf_{i \in I} \|x_i - \xi\| ,$$

and

$$\Gamma_2 : (\mathbb{R}^n)^m \to \mathbb{R}$$
$$(x_1, \ldots, x_m) \mapsto \Gamma_2(x) := \sup \left[\sup_{\xi \in \Omega_i} \|x_i - \xi\| \mid i \in I\right],$$

where

$$\Omega_i = D_i(x) \cap \Omega ,$$

then

$$\Gamma_1 = \Gamma_2 .$$

Emanating the preceding arguments the function value of a finite ball covering problem is the supremum of solutions to a finite number of convex optimization problems. Note, if $x \in \Omega$ then $\Omega_i \neq \emptyset$.

Lemma 4.3: *Given a compact set $\Omega \subset \mathbb{R}^n$ a finite set of positive integers $I := \{1, \ldots, m\}$ I, $\Omega \neq \emptyset$ and a fixed compact partition T of Ω i.e.*

$$\bigcup_{i \in I} T_i = \Omega , \quad T_i, \text{ nonempty and compact,}$$

the function

$$\Gamma^T : (\mathbb{R}^n)^m \to \mathbb{R}$$
$$(x_1, \ldots, x_m) \mapsto \Gamma^T(x) := \sup_{i \in I} \sigma_i^T(x) ,$$

where

$$\sigma_i^T : (\mathbb{R}^n)^m \to \mathbb{R}$$
$$(x_1, \ldots, x_m) \mapsto \sigma_i^T(x) := \sup_{\xi \in T_i} \|x_i - \xi\| .$$

is quasi-differentiable.

Proof: Due to Lemma 4.2 the function values of σ_i^T are equivalent to the supremum of a convex function over a convex compact set let us say C. Therefore σ_i^T is convex for every $i \in I$, since for every $x, y \in (\mathbb{R}^n)^m$ and for every $\lambda \in [0, 1]$

$$\begin{aligned} \sigma_i^T(\lambda \cdot x + (1-\lambda) y) &:= \sup_{\xi \in C} \|\lambda \cdot x_i + (1-\lambda) \cdot y_i - \xi\| \\ &:= \|\lambda \cdot x_i + (1-\lambda) \cdot y_i - \xi_0\| \\ &= \|\lambda \cdot x_i + (1-\lambda) \cdot y_i - (1-\lambda) \cdot \xi_0 - \lambda \cdot \xi_0\| \\ &\leqq \lambda \cdot \|x_i - \xi_0\| + (1-\lambda) \cdot \|y_i - \xi_0\| \\ &\leqq \lambda \cdot \sup_{\xi \in C} \|x_i - \xi\| + (1-\lambda) \cdot \sup_{\xi \in C} \|y_i - \xi\| . \end{aligned}$$

Convexity implies directional differentiability and as proposed in [2, page 122f.] the directional derivatives are given by

$$\frac{\partial \sigma_i^T(x)}{\partial g} = \sup_{v \in \underline{\partial} \sigma_i^T(x)} \langle v, g \rangle ,$$

where the quasi-differentials of σ_i^T at x are given by

$$D\sigma_i^T = [\{v \in (\mathbb{R}^n)^m \mid \sigma_i^T(y) \geqq \sigma_i^T(x) + \langle v, y - x \rangle \ \ \forall y \in (\mathbb{R}^n)^m\}, \{0\}] .$$

As stated in Theorem 2.0 the class of quasi-differentiable functions is closed with respect to taking the pointwise supremum of a finite number of quasi-differentiable functions and we conclude that Γ^T is quasi-differentiable and convex.

Applying Theorem 1.0 we obtain

$$D\Gamma^T(x):=[\text{conv}\,\{\underline{\partial}\sigma_k^T(x) \mid k\in R(x)\},\,\{0\}]\,,$$

which completes the proof. ■

Note that the sub-differentials of σ_i^T are the sets of all sub-gradients of σ_i^T at x as defined in [8, page 213ff.] and therefore convex and compact.

The following theorem states a possibility to generate a monotonously decreasing sequence of function values to a finite ball covering problem.

Theorem 4.2: *Given a compact set $\Omega\subset\mathbb{R}^n$, a finite set of positive integers $I:=\{1,\ldots,m\}$, I, $\Omega\neq\emptyset$ and a function*

$$\Gamma:(\mathbb{R}^n)^m\to\mathbb{R}$$
$$(x_1,\ldots,x_m)\mapsto\Gamma(x):=\sup_{\xi\in\Omega}\inf_{i\in I}\|z-\xi\|\,,$$

then $\Gamma(y)\leqq\Gamma(x)$ if y_i is chosen as

$$y_i:=\underset{z\in\Omega}{\text{arginf}}\sup_{\xi\in\Omega(x_i)}\|y_i-\xi\|\,.$$

Proof: For simplification let $X_i=\Omega(x_i)$ and $Y_i=\Omega(y_i)$. We have to show that the following estimation holds.

$$\begin{aligned}\Gamma(x)&:=\sup_{\xi\in\Omega}\inf_{i\in I}\|x_i-\xi\|\\&=\sup\,[\sup_{\xi\in X_i}\|x_i-\xi\|\mid i\in I]\\&\geqq\sup\,[\sup_{\xi\in X_i}\|y_i-\xi\|\mid i\in I]\\&\geqq\sup\,[\sup_{\xi\in Y_i}\|y_i-\xi\|\mid i\in I]\\&=\sup_{\xi\in\Omega}\inf_{i\in I}\|y_i-\xi\|:=\Gamma(y)\,.\end{aligned}$$

The equality relations hold due to Lemma 4.1. The first inequality is a consequence of the definition of $y:=(y_1,\ldots,y_m)$. To verify the second observe that

$$X_i\subseteqq\text{conv}\,(X_i)\subseteqq B\,(y_i,\,\sup_{\xi\in X_i}\|y_i-\xi\|)$$

and therefore

$$\Omega=\bigcup_{i\in I}X_i\subseteqq\bigcup_{i\in I}B(y_i,\,\sup_{\xi\in X_i}\|y_i-\xi\|)\,,$$

which implies that $\{B(y_i,\,\sup_{\xi\in X_i}\|y_i-\xi\|)\}_{i\in I}$ is a ball covering of Ω subject to y. Due to Lemma 3.0 however, $\Gamma(y)$ is a minimal covering subject to y. Hence the second inequality relation holds as well, which completes the proof. ■

In the sequel we will specify an algorithm to solve a class of finite ball covering problems on the plane.

5. An Algorithm to Solve a Restricted Finite Ball Covering Problem on A Class of Compact Subsets of $\mathbb{R}^2$

We will assume that $\Omega \subset \mathbb{R}^2$ can be described as the union of a finite number of nonempty convex polyhedrons. A feasible solution $x^0 = (x_1^0, \ldots, x_m^0) \in \Omega$ is known and for $i \neq j$, $x_i \neq x_j$ holds.

Algorithm II

Initialization: Let $k=0$, $I=\{1, \ldots, m\}$ and let $\varepsilon > 0$ be sufficiently small.

Step 1: Compute the DIRICHLET tessellation of Ω generated by x_k i.e. the set

$$\{\Omega_1^k, \ldots, \Omega_m^k\} := \{\Omega(x_1^k), \ldots, \Omega(x_m^k)\}.$$

Step 2: For every $i \in I$, $\Omega_i^k \neq \emptyset$ compute y_i^k, where

$$y_i^k := \operatorname*{arginf}_{z \in \Omega_i^k} \sup_{\xi \in \mathrm{ext}(\mathrm{conv}(\Omega_i^k))} \|z - \xi\| . \tag{5.0}$$

If $\Omega_i^k = \emptyset$ then $y_i^k = x_i^k$.

If the stopping relation holds i.e. if

$$\sum_{i=1}^{m} \|y_i^k - x_i^k\|^2 \leqq \varepsilon^2$$

then continue at Step 4.

Step 3: Let $x^{k+1} = (y_1^k, \ldots, y_m^k)$.

$k = k+1$.

Continue at Step 1.

Step 4:

$$x^{\mathrm{opt}} = (y_1^k, \ldots, y_m^k).$$

$$\Omega(x^{\mathrm{opt}}) := \sup_{i \in I} \sup_{\xi \in \Omega_i^k} \|x_i^{\mathrm{opt}} - \xi\| .$$

Remark to Step 1: The DIRICHLET tessellation can be computed for instance by a modified version of an algorithm suggested in [1] or [5].

Remark to Step 2: Under the given assumptions the sets $\mathrm{ext}(\mathrm{conv}(\Omega_i^k))$ are finite which consequences that the determination of steepest descent directions to the optimization problem (5.0) can easily be derived from Theorem 1.0 and Lemma 1.0. This problem can therefore be solved by the gradient type method described in Chapter 2. See also Step 2 of the example in the next chapter.

Stopping rule: Due to Theorem 4.2 the sequence $\{\Gamma|_{x_k}\}_{k \in \mathbb{N}}$ is only monotonuosly decreasing. Therefore we suggest to stop the algorithm if the sum of changes in the argument x does not exceed a given ε.

Corollary 5.0: *The sequence $\{\Gamma|_{x^k}\}_{k\in\mathbb{N}}$ generated by Algorithm II is converging.*

Remark 5.0: Algorithm II computes stationary points.

6. An Example

We will use simple graphs to represent convex polyhedrons in the following way:

Assume that a set M is a convex polyhedron with a set of extremal points $\text{ext}(M)=\{m_1, \ldots, m_p\}$, $p\in\mathbb{N}$, $p\geqq 2$.

Then M can be represented by the simple graph $G(M):=[V, E, \alpha]$, where

$$V:=\{v_1, \ldots, v_p\},$$
$$\alpha: V\to\text{ext}(M)\subset\mathbb{R}^2$$
$$v\mapsto\alpha(v_i):=m_i, \quad \text{and}$$
$$E:=\{e_1, \ldots, e_p\}:=\{(v_i, v_j)=(v_j, v_i) \mid [\alpha(v_i), \alpha(v_j)]\subseteqq\text{bd}(M)\},$$

respectively denote the set of vertices, a bijective map of V into $\text{ext}(M)$ and the set of edges.

Let ε be a sufficiently small positive real number and consider the graphs $G^t:=[V^t, E^t, \alpha^t]$, $t\in T:=\{1, 2, 3\}$,

$$V^t:=\{v_1^t, v_2^t, v_3^t, v_4^t\}, \quad E^t:=\{(v_1^t, v_2^t), (v_2^t, v_3^t), (v_3^t, v_4^t), (v_4^t, v_1^t)\}, \quad t\in T$$

$$\alpha^1(v_1^1)=(0.0, 0.0)^\top, \quad \alpha^2(v_1^2)=(0.0, 6.0)^\top, \quad \alpha^3(v_1^3)=(2.0, 2.0)^\top$$
$$\alpha^1(v_2^1)=(0.0, 6.0)^\top, \quad \alpha^2(v_2^2)=(5.0, 6.0)^\top, \quad \alpha^3(v_2^3)=(8.5, 2.0)^\top$$
$$\alpha^1(v_3^1)=(2.0, 4.0)^\top, \quad \alpha^2(v_3^2)=(5.0, 4.0)^\top, \quad \alpha^3(v_3^3)=(8.5, 0.0)^\top$$
$$\alpha^1(v_4^1)=(2.0, 2.0)^\top, \quad \alpha^2(v_4^2)=(2.0, 4.0)^\top, \quad \alpha^3(v_4^3)=(0.0, 0.0)^\top$$

We start with $x^0:=(x_1^0, x_2^0, x_3^0)=\left(\begin{pmatrix}1.0\\0.5\end{pmatrix}, \begin{pmatrix}5.0\\0.5\end{pmatrix}, \begin{pmatrix}6.0\\1.5\end{pmatrix}\right)$.

Step 1: The tessellation $D(x^0)=\{D_1(x^0), D_2(x^0), D_3(x^0)\}$ is represented by:

1. $G_1^1:=[V_1^1, E_1^1, \alpha_1^1]$, $\quad G_1^2:=[V_1^2, E_1^2, \alpha_1^2]$, $\quad G_1^3:=[V_1^3, E_1^3, \alpha_1^3]$

$$\alpha_1^1(v_1)=(0.0, 0.0)^\top, \quad \alpha_1^2(v_1)=(0.0, 6.0)^\top, \quad \alpha_1^3(v_1)=(2.0, 2.0)^\top,$$
$$\alpha_1^1(v_2)=(0.0, 6.0)^\top, \quad \alpha_1^2(v_2)=(2.5, 6.0)^\top, \quad \alpha_1^3(v_2)=(3.0, 2.0)^\top,$$
$$\alpha_1^1(v_3)=(2.0, 4.0)^\top, \quad \alpha_1^2(v_3)=(2.9, 4.0)^\top, \quad \alpha_1^3(v_3)=(3.0, 0.0)^\top,$$
$$\alpha_1^1(v_4)=(2.0, 2.0)^\top, \quad \alpha_1^2(v_4)=(2.0, 4.0)^\top, \quad \alpha_1^3(v_4)=(0.0, 0.0)^\top,$$

2. $G_2^1:=[V_2^1, E_2^1, \alpha_2^1]$, $\quad G_2^2:=[V_2^2, E_2^2, \alpha_2^2]$, $\quad G_2^3:=[V_2^3, E_2^3, \alpha_2^3]$,

$$\alpha_2^3(v_1)=(3.0, 2.0)^\top,$$
$$V_2^1=\emptyset \qquad V_2^2=\emptyset \qquad \alpha_2^3(v_2)=(4.5, 2.0)^\top,$$
$$\alpha_2^3(v_3)=(6.5, 0.0)^\top,$$
$$\alpha_2^3(v_4)=(3.0, 0.0)^\top,$$

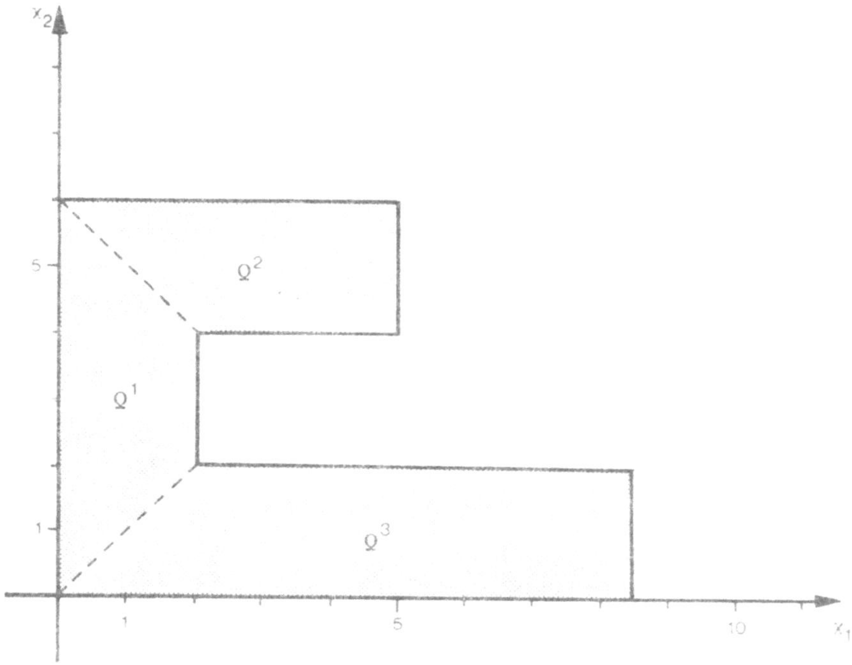

Fig. 6.0: The set $\Omega := \bigcup_{t \in T} \Omega^t = \bigcup_{t \in T} \text{conv}\,(\alpha(V^t))$.

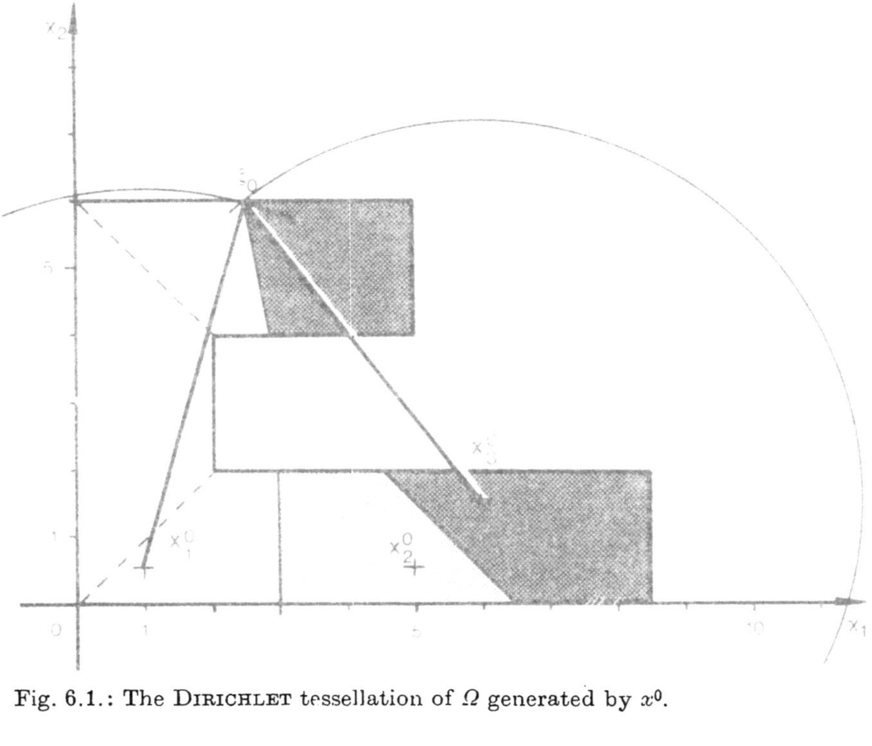

Fig. 6.1.: The DIRICHLET tessellation of Ω generated by x^0.

3. $G_3^1 := [V_3^1, E_3^1, \alpha_3^1]$, $G_3^2 := [V_3^2, E_3^2, \alpha_3^2]$, $G_3^3 := [V_3^3, E_3^3, \alpha_3^3]$,

$$V_3^1 = \varnothing \qquad \begin{array}{ll} \alpha_3^2(v_1) = (2.5,\ 6.0)^{\mathsf T}, & \alpha_3^3(v_1) = (4{,}5,\ 2.0)^{\mathsf T}, \\ \alpha_3^2(v_2) = (5.0,\ 6.0)^{\mathsf T}, & \alpha_3^3(v_2) = (8.5,\ 2.0)^{\mathsf T}, \\ \alpha_3^2(v_3) = (5.0,\ 4.0)^{\mathsf T}, & \alpha_3^3(v_3) = (8.5,\ 0.0)^{\mathsf T}, \\ \alpha_3^2(v_4) = (2.9,\ 4.0)^{\mathsf T}, & \alpha_3^3(v_4) = (6.5,\ 0.0)^{\mathsf T}. \end{array}$$

The VORONOI sets of Ω generated by x^0 are then given by

$$D_i(x^0) = \bigcup_{t \in T} \operatorname{conv}(\alpha(V_i^t)),$$

Step 2: Computation of a descent:

For all $(i, t) \in I \times T$, $V_i^t \neq \varnothing$ we have to find the minimum to the optimization problem:

Minimize

$$f_i^t \mathbb{R}^n \to \mathbb{R} \tag{6.0}$$

$$(y_1, y_2) \mapsto f_i^t(y) := \sup_{\xi \in \bigcup_{t \in T} \{\alpha(V_i^t)\}} \|y - \xi\|,$$

subject to

$$(y_1, y_2) \in M_i^t := \operatorname{conv}(\alpha(V_i^t)).$$

For simplification let $U_i := \bigcup_{t \in T} \{\alpha(V_i^t)\}$.

Let us compute a steepest descent direction of f_3^3 at $y^0 = x_3^0$.

$$f_3^3 : \mathbb{R}^n \to \mathbb{R}$$

$$(y_1, y_2) \mapsto \sup_{\xi \in U_3} \|y - \xi\|..$$

subject to

$$y \in M_3^3 = \operatorname{conv}\left(\begin{pmatrix}4.5\\2.0\end{pmatrix}\begin{pmatrix}8.5\\2.0\end{pmatrix}\begin{pmatrix}8.5\\0.0\end{pmatrix}\begin{pmatrix}6.5\\0.0\end{pmatrix}\right)$$

$$U_3 = \left(\begin{pmatrix}2.5\\6.0\end{pmatrix}\begin{pmatrix}5.0\\6.0\end{pmatrix}\begin{pmatrix}5.0\\4.0\end{pmatrix}\begin{pmatrix}2.9\\4.0\end{pmatrix}\begin{pmatrix}4.5\\2.0\end{pmatrix}\begin{pmatrix}8.5\\2.0\end{pmatrix}\begin{pmatrix}8.5\\0.0\end{pmatrix}\begin{pmatrix}6.5\\0.0\end{pmatrix}\right).$$

Applying Theorem 1.0 we obtain

$$Df_3^3\,|_{y^0} = \left(\begin{pmatrix}0.6139\\-0.7893\end{pmatrix}\begin{pmatrix}0.0000\\0.0000\end{pmatrix}\right).$$

This is not surprising since at y^0 there is only one $\xi_0 \in U_3$ to take the supremum, i.e. $\xi_0 = (2.5,\ 6.0)^{\mathsf T}$. Since $y^0 \neq \xi_0$ we find that f_3^3 is continuously differentiable at y^0 and $\partial f_3^3\,|_{y^0}$ reduces to $\nabla f_3^3\,|_{y^0}$. A steepest descent direction of f_3^3 at y^0 is then given by

$$g^*\,|_{y^0} = -\nabla f_3^3\,|_{y^0} = \begin{pmatrix}-0.6139\\0.7893\end{pmatrix}.$$

$y^1 := (5.6111,\ 2.0000)^{\mathsf T}$ is obtained by a line search.

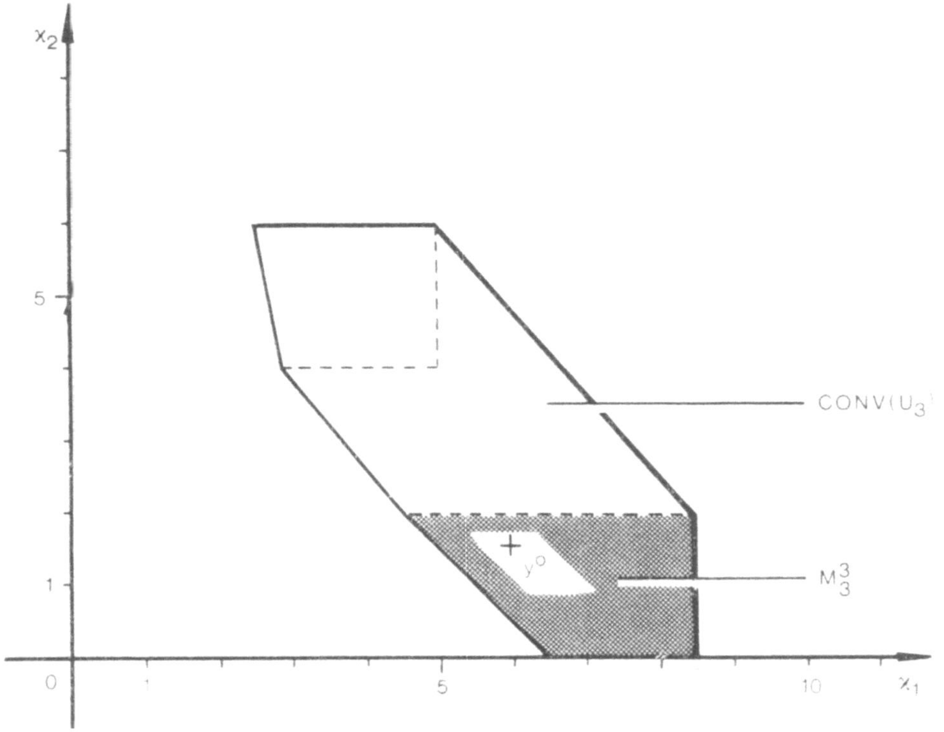

Fig. 6.2: The optimization problem min $\{f_3^3(y) | y \in M_3^3\}$.

Since $y^1 \in \text{bd}(M_3^3)$ we replace f_3^3 by

$$f_{3_{y^1}}^3 : \mathbb{R}^2 \to \mathbb{R}$$

$$(z_1, z_2) \mapsto \sup\{g_1(z), g_2(z)\},$$

where

$$g_1, g_2 : \mathbb{R}^2 \to \mathbb{R}$$

$$g_1(z) := \sup_{\xi \in U_3} \|z - \xi\| - f_3^3 |_{y^1},$$

$$g_2(z) := \sup (z_2 - 2,\ z_1 - 8.5,\ -z_2,\ -z_1 - z_2 + 6.5).$$

Note that g_2 can easily be derived from $\alpha(V_3^3)$ and E_3^3.
The quasidifferential of $f_{3_{y^1}}^3$ at y^1 is given by

$$Df_{3_{y^1}}^3 |_{y^1} = \left(\text{conv}\left\{\begin{pmatrix} 0.6139 \\ -0.7894 \end{pmatrix}, \begin{pmatrix} -1.0000 \\ 0.0000 \end{pmatrix}\right\}, \left\{\begin{pmatrix} 0.0000 \\ 0.0000 \end{pmatrix}\right\}\right).$$

As a steepest descent direction of $f_{3y^1}^3$ at y^1 we obtain

$$g^* |_{y^1} = \begin{pmatrix} -0.4394 \\ -0.8983 \end{pmatrix}.$$

We stop the iteration at this point. The minimum is taken at

$$y^* = (4.5000,\ 2.0000)^\mathsf{T}.$$

Having evaluated the solutions to the subproblems (6.0) we obtain

$$x^1 = (x_1^1, x_2^1, x_3^1) = \left(\begin{pmatrix} 1.5000 \\ 3.0000 \end{pmatrix} \begin{pmatrix} 4.7500 \\ 1.0000 \end{pmatrix} \begin{pmatrix} 4.5000 \\ 2.0000 \end{pmatrix} \right).$$

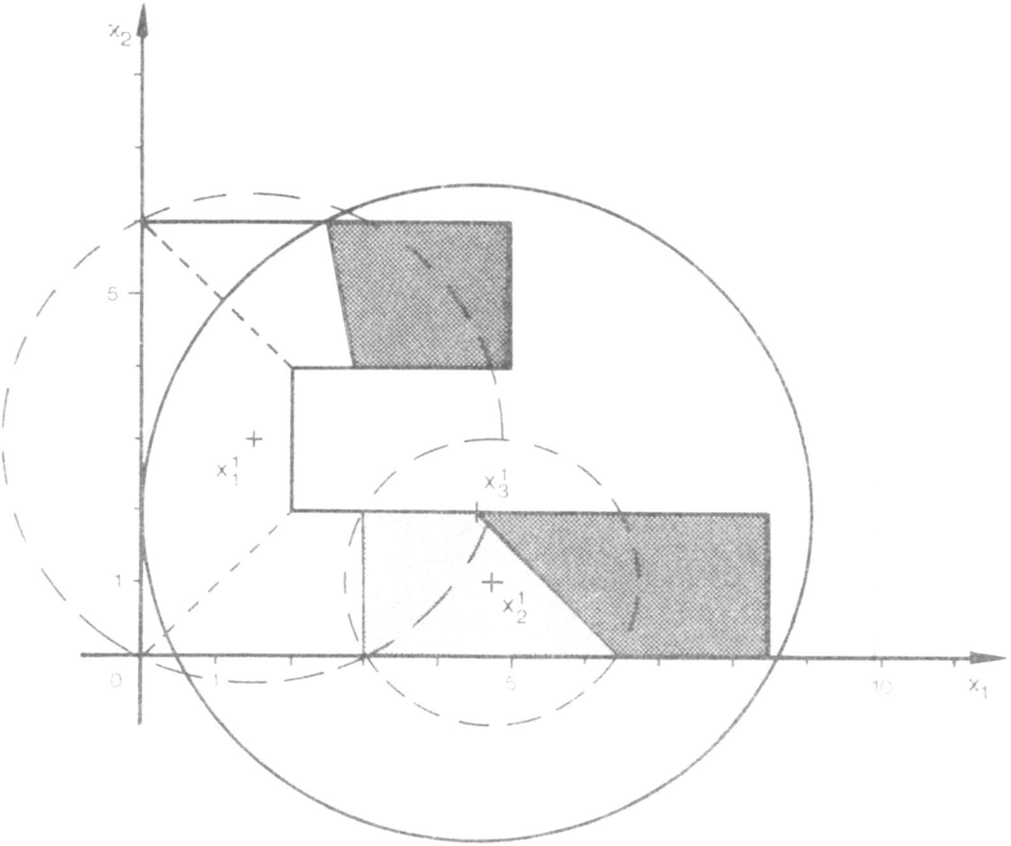

Fig. 6.3: Minimal coverings of the VORONOI sets generated by x^0.

The DIRICHLET tesselation of the second iteration leads to the situation of figure 6.4.

Due to the lack of space we stop at this point. More details (DIRICHLET tesselations, implementation hints etc.) may be found in the extended version of this paper which is going to appear in the series "Collaborative Paper" at the International Institute for Applied Systems Analysis in A-2361 Laxenburg, Austria.

An optimal solution of the discussed example is shown in figure 6.5.

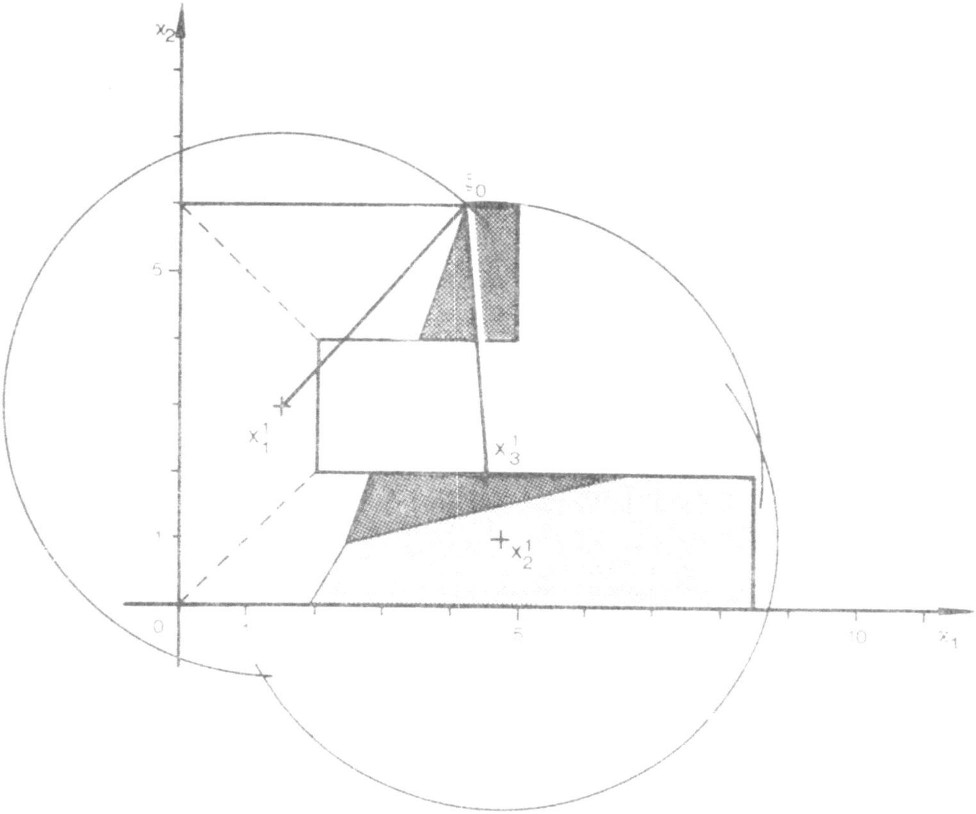

Fig. 6.4: The DIRICHLET tessellation of Ω generated by x^1.

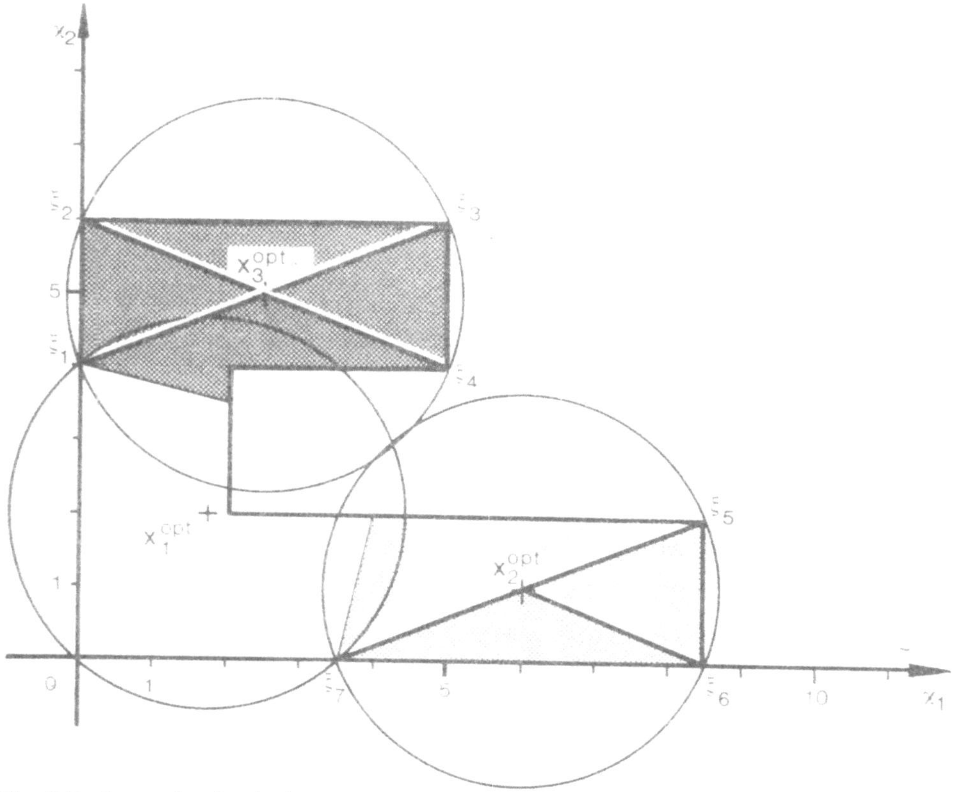

Fig. 6.5: An optimal solution.

Final Remark

The example was well arranged to demonstrate the basic idea of our algorithm. In general with increasing number of iterations however, the quasi-differentials get more complicated, depending on the number of elements in Ω, where the function value of the objective is attained.

References

[1] BOWYER, A.: Computing DIRICHLET Tessellations. The Computer Journal. London, **24** (1981) 2, 162–166.

[2] DEMYANOV, V. F.; A. M. RUBINOV: Quasidifferential Calculus. Optimization Software, Springer-Verlag, Berlin, Heidelberg, New York 1986.

[3] DEMYANOV, V. F.; L. V. VASILEV: Nondifferentiable Optimization. Optimization Software, Springer-Verlag, Berlin, Heidelberg, New York 1985.

[4] DEMYANOV, V. F.; I. S. ZABRODIN: Directional Differentiability of a Continual Maximum Function of Quasi-differentiable Functions. WP-83-58 IIASA, Laxenburg 1983.

[5] GREEN, P. J.; R. SIBSON: Computing DIRICHLET Tessellations in the Plane. The Computer Journal, London, **21** (1978) 2, 168–173.

[6] PALLASCHKE, D.; P. RECHT: On the Steepest Descent Method for a Class of Quasidifferentiable Optimization Problems. Nondifferentiable Optimization: Motivations and Applications Proceedings, Sopron, Hungary, 1984, pp. 252–263, Springer-Verlag Berlin, Heidelberg, New York, Tokyo 1985.

[7] PALLASCHKE, D.; P. RECHT: Urbanski, R., On Locally-LIPSCHITZ Quasi-Differentiable Functions in BANACH-Spaces. Optimization, **17** (1986) 287–295.

[8] ROCKAFELLAR, R. T.: Convex Analysis. Princeton University Press, Princeton 1970

Received February 1987, revised December 1987

H. Jandl
Inst. f. Statistik und math.
Wirtschaftstheorie der
Universität Karlsruhe
Kollegium am Schloß Bau III
D - 7500 Karlsruhe 1

K. Wieder
Inst. f. Statistik und math.
Wirtschaftstheorie der
Universität Karlsruhe
Kollegium am Schloß Bau III
D - 7500 Karlsruhe 1

optimization 19 (1988) 6, 803–817

A Cutting–Plane Method for Quadratic Semi-Infinite Programming Problems

R. Tichatschke and V. Nebeling

Pädagogische Hochschule "Liselotte Herrmann" Güstrow
Pädagogische Hochschule "N. K. Krupskaja" Halle

Summary: A cutting plane algorithm for solving convex quadratic semi-infinite programming problems is presented. Nonbinding constraints can be dropped. Its arithmetic convergence rate is proved by taking into consideration the error of the approximate solution of the auxiliary problem to calculate the most violate constraint. An implementable variant of this method is described which is due to the adaptive discretization of the index set and its stability is shown. Computational experiments show the behaviour of the method.

AMS 1980 Subject Classification:

Primary: 49 D 39

Key words: Numerical methods, cutting-plane methods, semi-infinite programs, rates of convergence, numerical stability.

1. Introduction

Semi-infinite programming deals with the problem

$$\min \{f(x) \mid g(x, t) \leqq 0 \quad t \in T,\, x \in X\}\,. \tag{1.1}$$

If T is a finite set then (1.1) is an ordinary finite program with a finite number of inequality constraints. But there are many problems in which the set T is infinite, e.g. problems with constraints depending on time- or spacecoordinates like in extremal problems of mathematical physics.

Because semi-infinite programming problems can not be solved directly, discretization methods must be applied. The idea of adaptive discretization is based on refining a grid $T_k \subset T$ in each iteration by using available information, which is necessary to determine an approximate solution. In this way, semi-infinite programming problems can be solved effectively.

Cutting-plane methods are special adaptive discretization methods and in the present paper such an approach is considered for quadratic semi-infinite problems. In this approach the subproblems can be solved in a finite number of iterations and an increase of the dimension is prevented by dropping nonbinding constraints. In this way the advantages of various methods given in the literature are combined (cf. Blankenship/Falk [2], Tichatschke/Lohse [20]).

A motivation to consider this class of problems is a number of methods for solving nonlinear semi-infinite problems which lead to quadratic subproblems. This is the similar case as the recursive quadratic approximation technique for finite nonlinear problems.

In Chapter 2 a conceptual cutting-plane method is described which is basic for the following. For this method convergence with an arithmetic rate is proved in Chapter 3. An implementable version of this method is developed and its stability is shown in Chapter 4. Finally, computational experiments are presented which give some information about the behaviour of the presented algorithm.

2. A Generalized Cutting-Plane Method

Problem (1.1)

$$\min \{f(x) \mid g(x, t) \leqq 0 \quad t \in T,\ x \in X\}$$

is considered under the following assumptions:

Assumption 2.1.:

(i) $X \subset \boldsymbol{R}^n$ *is a compact, convex and nonempty set*;

(ii) $T \subset \boldsymbol{R}^p$ *compact and nonempty*;

(iii) $g: \boldsymbol{R}^p \times \boldsymbol{R}^n \to \boldsymbol{R}$, $g(x, t)$ *continuous on* $X \times T$, $g(\cdot, t)$ *convex for all* t *and differentiable on* X *and* $\nabla_x g(x, t)$ *continuous on* $X \times T$;

(iv) $\exists \hat{x} \in \operatorname{int} X: g(\hat{x}, t) < 0$ *for all* T *(without loss of generality we assume that* $\hat{x}$ *is not optimal)*;

(v) $f: \boldsymbol{R}^n \to \boldsymbol{R}$, $f(x) = 1/2 x^{\mathsf{T}} C x - p^{\mathsf{T}} x$, C *positive definite.*

Let the set X be described by a finite number of convex constraints

$$X := \{x \in \boldsymbol{R}^n \mid f_j(x) \leqq 0,\ f_j \text{ convex},\ j \in J\},$$
$$J := \{1, 2, \ldots, q\}.$$

This case requires to determine which of the inequalities $f_j(x) \leqq 0$ are violated by the current solution and to linearize f_j in the same way as the constraints $g(x, t) \leqq 0$, $t \in T$. This is possible without fundamental difficulties (cf. GONZAGA/POLAK [6]). For the sake of an easy and clear representation only the case of affin linear constraints $f_j(x) = A_j x - b_j$, $j \in J$ is considered, where A_j denotes the j-th row of a (q, n)-matrix and b_j is the j-th component of the vector $b \in \boldsymbol{R}^q$.

Hence problem (1.1) can be stated as follows

$$\min \{1/2\, x^{\mathsf{T}} C x - p^{\mathsf{T}} x \mid g(x, t) \leqq 0 \quad t \in T,\ A_j x \leqq b_j\ j \in J\}. \tag{2.1}$$

Problem (2.1) is treated by a cutting-plane method. In each iteration a cut is generated by means of the current solution, which improves the outer approximation of the feasible region of problem (2.1). To generate cuts those considered by KELLEY [12] it is necessary to find the most violated constraint in each iteration by solving the following auxiliary problem:

(H_k) $\quad \max \{g(x^k, t) \mid t \in T\}, \quad x^k$ a current solution.

That means, a global maximum of problem (H_k) is required. The auxiliary problem (H_k) is a nonlinear (i.e. nonconvex, nondifferentiable) programming problem. It is treated by the following discretization: Let $\{T_m\}$ denote a set sequence and it is assumed that $T_m \to T$ in the sense of the HAUSDORFF distance. Now, one can replace (H_k) by a sequence of approximate problems $\{(H_k^m)\}$, $m=1, 2, \ldots$

(H_k^m) $\quad \max\{g(x^k, t) \mid t \in T_m\}$.

The convergence of the optimal values and solutions of the sequence $\{(H_k^m)\}$ to those of (H_k) is secured by Assumption 2.1. (cf. TICHATSCHKE [19]). The iteration is stopped if the feasible region T_m of (H_k^m) is a sufficiently accurate approximation of the set T.

We start by giving a conceptual algorithm:

Algorithm A:

$k:=1$, $T^0:=\emptyset$, $Z^0:=X$ and choose T_m for $m=1, 2, \ldots$ such that $T_m \subset T$, $T_m \to T$;

S1: *Compute a solution of the subproblem*

(P_{k-1}) $\quad \min\{f(x) \mid x \in Z^{k-1}\}$.

S2: *Compute an exact solution t^m of the auxiliary problem*

(H_k^m) $\quad \max\{g(x^k, t) \mid t \in T_m\}$.

S3a: *If $g(x^k, t^m) < \max\{g(x^k, t) \mid t \in T)$ return to* S2 *with $m:=m+1$, otherwise set $t^k:=t^m$.*

S3b: *If $g(x^k, t^k) > 0$ go to* S4, *otherwise stop, x^k is an approximate solution of* (2.1).

S4: *Add the cut*

$$s_k(x^k, t^k; x) := g(x^k, t^k) + \nabla_x g(x^k, t^k)^{\mathsf{T}} (x - x^k) \leqq 0$$

and determine

$$F^k := \{t^j \in T^{k-1} \mid s_j(x^j, t^j; x^k) = 0\},$$

where T^{k-1} is the index set of (P_{k-1}) and set

$$T^k := F^k \cup \{t^k\},$$

$$Z^k := \bigcap_{t^j \in T^k} \{x \mid s_j(x^j, t^j; x) \leqq 0\} \cap X .$$

Return to **S1** *with $k:=k+1$.*

Remark 2.2.: (i) Of course S3 is not implementable. Using linear cuts, the feasible region Z^{k-1} of subproblem (P_{k-1}) will remain polyedral, i.e. (P_{k-1}) is a quadratic programming problem with a finite number of linear constraints.

(ii) In S4—with respect to the current iterate x^k—the nonbinding constraints are deleted. It is easily seen that both points x^k and x^{k+1} belong to the set Z^{k-1}, but x^k does not belong to Z^k.

Lemma 2.3.: *The cut $s_k(x^k, t^k; x) \leqq 0$ is active in the point x^{k+1}, i.e.*

$$s_k(x^k, t^k; x^{k+1}) = 0 \quad \textit{for all} \quad k .$$

Proof: According to S4 of Algorithm we have

$$F^k = \{t^j \in T^{k-1} \mid s_j(x^j, t^j; x^k) = 0\}$$

and $T^k = F^k \cup \{t^k\}$. Assume that $s_k(x^k, t^k; x^{k+1}) < 0$, then there exsits a $t^j \in T^k$ with $s_j(x^k, t^j; x^{k+1}) = 0$, otherwise the cut is ineffectual. Because the cut s_k especially cuts off the point x^k from Z^{k-1} there must be a point $z \in Z^{k-1}$ with $z \in (x^k, x^{k+1}]$ and $s_k(x^k, t^k; z) = 0$. Because x^k is a solution of the subproblem (P_{k-1}) one has $f(z) \geqq f(x^k)$.

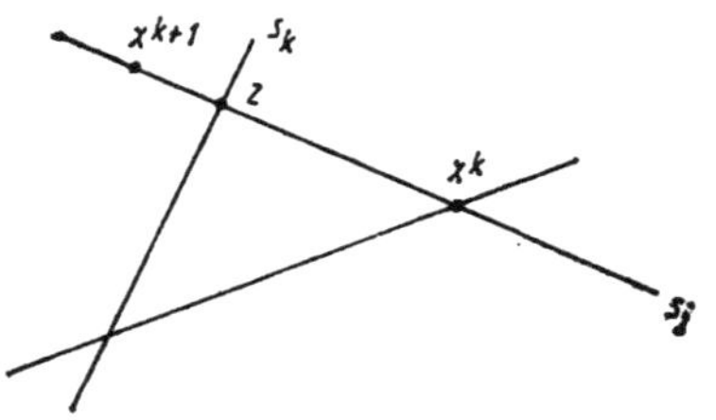

Fig. 2.1.

On the other hand one has also $z \in Z^k$ and $f(z) \geqq f(x^{k+1})$ by the same arguments. This is a contradiction to the strict convexity of f. ∎

3. Convergence Properties

The auxiliary problem (H_k) of Algorithm A is solved by a sequence of appropriate problems (H_k^m). If in S3a the iteration stops with $m = m_0$, one obtains an approximate solution $t^k = t^{m_0}$ of the global maximum at t^{k^*} of (H_k).

Let d_k be the deviation of the value of the approximately computed, most violated constraint $g(x, t^k)$ of $(H_k^{m_0})$ at the point x^k to the optimal value

$$g(x^k, t^{k^*}) := \max \{g(x^k, t) \mid t \in T\} \quad \text{of} \quad (H_k)$$

$$d_k := g(x^k, t^{k^*}) - g(x^k, t^k)\,. \tag{3.1}$$

Assumption 3.1.: *For the sequence* $\{d_k\}$ $\lim_{k\to\infty} d_k = 0$ *is valid.*

Lemma 3.2.: *Under Assumptions* 2.1. *and* 3.1. *the sequence* $\{x^k\}$ *let be generated by the Algorithm* A. *If* $\{x^k\}$ *converges to a point* $\bar{x}$ *then* $\bar{x}$ *is a solution of Problem* (2.1).

Proof: It is to show that the limit point $\bar{x}$ of the sequence $\{x^k\}$ is feasible and

$$f(\bar{x}) = f^* = \min \{1/2 x^\mathsf{T} C x - p^\mathsf{T} x \mid g(x, t) \leqq 0 \quad t \in T,\ Ax \leqq b\}\,.$$

Let $\bar{x}$ be not feasible to Problem (2.1). By the assumption of continuity of g and compactness of X and T there exists for a sufficiently large m a point $t^0 \in T_m$ with $g(\bar{x}, t^0) > 0$, i.e., there is a sufficiently large k_0 such that

$$g(x^k, t^0) > -d_k \quad \text{if} \quad k \geqq k_0\,.$$

(By S3 x^k is feasible if $g(x^k, t^0) \leqq -d_k$).

Let $\{x^{k_j}\}$ be a subsequence such that the corresponding sequence $\{t^{k_j}\}$ converges to $\hat{t} \in T$. Because t^{k_j} is a solution of (H_{k_j}) the following estimations hold:

$$g(x^{k_j}, t^{k_j}) \geqq g(x^{k_j}, t^0) > -d_{k_j} .$$

Taking the limit it follows

$$g(\bar{x}, \hat{t}) \geqq g(\bar{x}, t^0) > 0 . \tag{3.2}$$

Because of Remark 2.2. (ii)

$$s_{k_j}(x^{k_j}, t^{k_j}; x^{k_j+1}) \leqq 0 \quad \text{holds}$$

and the solution x^{k_j+1} of (P_{k_j}) fulfils the cut s_{k_j}. Therefore, taking the limit, one obtains

$$s(\bar{x}, \hat{t}; \bar{x}) = g(\bar{x}, \hat{t}) + \nabla_x g(\bar{x}, \hat{t})^{\mathsf{T}} (\bar{x} - \bar{x}) = g(\bar{x}, \hat{t}) \leqq 0 .$$

This is a contradiction to (3.2), thus $\bar{x}$ is a feasible point to problem (2.1) and $f(\bar{x}) \geqq f^*$. On the other hand for the sequence $\{f(x^k)\}$ holds $f(x^k) \leqq f^*$. The desired conclusion follows for $k \to \infty$. ■

The question of constraint dropping is essential for cutting-plane methods. However, to ensure convergence of these methods, the objective function has to satisfy some conditions (cf. TOPKIS [23], EAVES/ZANGWILL [4], TICHATSCHKE/LOHSE [20]). The basic idea for such results is that nonbinding constraints can only be dropped if the values of the objective function of the iterates do not stagnate.

Under the above assumptions one can prove the following statement:

Lemma 3.3.: *Let the Assumptions* 2.1. *and* 3.1. *be satisfied and let the sequence* $\{x^k\}$ *be generated by the Algorithm* A. *Then the sequence of the corresponding values of the objective function* $\{f(x^k)\}$ *is strictly monoton increasing.*

Proof: Let be assumed that $f(x^{k+1}) \leqq f(x^k)$ for some k. One has for every $t^j \in T^{k-1} \setminus F^k$ by S4 of Algorithm A

$$s_j(x^k, t^j; x^k) < 0 ,$$

so that for $\alpha \in [0, 1]$ sufficiently close to 1 holds

$$s_j (x^k, t^j; \alpha x^k + (1 - \alpha) x^{k+1}) < 0 .$$

Thus, there exists some $\bar{\alpha} \in (0, 1]$ such that

$$s_j \left(x^k, t^j; \bar{\alpha} x^k + (1 - \bar{\alpha}) x^{k+1}\right) < 0 \quad \forall t^j \in T^{k-1} \setminus F^k .$$

Therefore

$$\bar{x}^k := \bar{\alpha} x^k + (1 - \bar{\alpha}) x^{k+1} \in Z^{k-1}$$

for the convex set Z^{k-1}, i.e., $\bar{x}^k$ is feasible for each k.

Because f is strictly convex it follows by definition

$$f(\bar{x}^k) < \bar{\alpha} f(x^k) + (1 - \bar{\alpha}) f(x^{k+1}) \leqq f(x^k) ,$$

which is a contradiction to

$$f(x^k) = \min \{f(x) \mid x \in Z^{k-1}\} . \quad ■$$

Using Lemma 3.2. the convergence of the sequence of iterates $\{x^k\}$ generated by Algorithm A can be proved.

Theorem 3.4.: *Under the assumptions of Lemma* 3.2. *the sequence* $\{x^k\}$ *converges to the unique solution of the problem* (2.1).

Proof: Let two subsequences of $\{x^k\}$ be given where $x^{k_j} \to \bar{x}$ and $x^{n_j} \to \bar{\bar{x}}$. Now, using Lemma 3.2., one can prove $f(\bar{x}) = f(\bar{\bar{x}}) = f^*$.

Let $\bar{x} \neq \bar{\bar{x}}$, then

$$f\left(\alpha \bar{x} + (1-\alpha)\,\bar{\bar{x}}\right) < \alpha f(\bar{x}) + (1-\alpha)\, f(\bar{\bar{x}}) = f^* \,, \tag{3.3}$$

since f is strictly convex. But $\bar{x}$, $\bar{\bar{x}}$ are feasible to problem (2.1), and the feasible region of (2.1) is convex, then by Lemma 3.2. it follows that $\alpha\bar{x} + (1-\alpha)\,\bar{\bar{x}}$ is also feasible to (2.1) and therefore (3.3) is a contradiction to the fact that $\bar{x}$ and $\bar{\bar{x}}$ are solutions of problem (2.1). ■

It is well known that cutting-plane methods for finite programming problems converge with an arithmetic convergence rate (cf. Topkis [24], Levitin/Poljak [14]). Analogous results can be shown for infinite programming problems (cf. Kapur [11], Oettli [15]). In the following we will show that an analogous convergence rate can be proved for Algorithm A.

Assumption 2.1. implies the uniform convexity of the objective function, i.e., for each x', $x'' \in X$ there exists a constant $\gamma > 0$ such that

$$f(1/2\ (x' + x'')) \leqq 1/2 f(x') + 1/2 f(x'') - \gamma\ \|x' - x''\|^2 \,. \tag{3.4}$$

Furthermore, Assumption 2.1. secures that there are two constants $N > 0$, $M > 0$ such that

$$\|\nabla f(x)\| \leqq M \quad \forall x \in X \quad \text{and} \quad \|\nabla_x g(x, t)\| \leqq N \quad \forall x \in X, \quad \forall t \in T \,.$$

Therefore, the following statement is valid:

Theorem 3.5.: *Let both Assumptions* 2.1. *and* 3.1. *be satisfied and the sequences* $\{x^k\}$ *and* $\{f(x^k)\}$ *be generated by Algorithm* A. *Then* $\{x^k\}$ *and* $\{f(x^k)\}$ *have both arithmetic convergence rates, i.e. there exist constants* k_0 *and* a_1, a_2 *such that*

$$|f^* - f(x^k)| \leqq a_1/k \quad \textit{for all} \quad k \geqq k_0$$

and

$$\|x^* - x^k\| \leqq a_2/\sqrt{k} \quad \textit{for all} \quad k \geqq k_0 \,,$$

where

$$a_1 = (1/\gamma)\left((MN) \cdot \|\hat{x} - x^k\| / (-(g\hat{x}, t^k) - d_0)\right)^2 ,$$
$$a_2 = (1/\gamma)\left((MN) \cdot \|\hat{x} - x^k\| / (-(g\hat{x}, t^k) - d_0)\right) .$$

Proof: Because of Assumption 2.1. (iv) for sufficiently small d_0 there exists some $\hat{x} \in X$ with

$$g(\hat{x}, t) < -d_0 \quad \text{for each} \quad t \in T \quad \text{and} \quad k \geqq k_0 \,.$$

But $g(x^k, t^k) \geqq 0$, there is some $\lambda_k \in (0, 1)$ such that

$$-d_0 = g\left(\lambda_k x^k + (1-\lambda_k)\,\hat{x},\, t^k\right) \leqq \lambda_k g(x^k, t^k) + (1-\lambda_k)\, g(\hat{x}, t^k)$$

which yields

$$g(x^k, t^k) \geqq ((1-\lambda_k)/\lambda_k)\,(-g(\hat{x}, t^k) - d_0) \quad \forall k \geqq k_0\,, \tag{3.5}$$

x^{k+1} is a solution of (P_k), hence

$$g(x^k, t^k) + \nabla_x g(x^k, t^k)^\mathsf{T}\,(x^{k+1} - x^k) \leqq 0,$$

so that with (3.5)

$$((1-\lambda_k)/\lambda_k)\,(-g(\hat{x}, t^k) - d_0) \leqq \nabla_x g(x^k, t^k)^\mathsf{T}\,(x^k - x^{k+1}) \leqq N\,\|x^k - x^{k+1}\|$$

and

$$1 - \lambda_k \leqq N\,\|x^k - x^{k+1}\|/(-g(\hat{x}, t^k) - d_0)\,. \tag{3.6}$$

Because $\lambda_k x^k + (1-\lambda_k)\,\hat{x}$ is a feasible point to problem (2.1)

$$f^* \leqq f\,(\lambda_k x^k + (1-\lambda_k)\,\hat{x})$$

and

$$f^* - f(x^k) \leqq (1-\lambda_k)\,(f(\hat{x}) - f(x^k)) \leqq (1-\lambda_k)\,\nabla f(\hat{x})^\mathsf{T}\,(\hat{x} - x^k)$$
$$\leqq M\,(1-\lambda_k)\,\|\hat{x} - x^k\|\,.$$

Together with (3.6) the estimation

$$\varepsilon_k := f^* - f(x^k) \leqq (MN)\,\|\hat{x} - x^k\| \cdot \|x^k - x^{k+1}\|/(-g(\hat{x}, t^k) - d_0)$$
$$= K_1\,\|x^k - x^{k+1}\| \tag{3.7}$$

is true.

f is strictly convex and therefore with (3.4) the following inequality is fulfilled

$$f(x^{k+1}) - f(x^k) \geqq \gamma \cdot \|x^k - x^{k+1}\|^2\,.$$

With $\varepsilon_k - \varepsilon_{k+1} = f(x^{k+1}) - f(x^k)$ one obtains by (3.7)

$$\varepsilon_k^2 \leqq K_1^2 \cdot \|x^k - x^{k+1}\|^2 \leqq (1/\gamma)\,(\varepsilon_k - \varepsilon_{k+1})\,K_1^2\,,$$

respectively

$$\varepsilon_{k+1} \leqq \varepsilon_k\,(1 - K_2\varepsilon_k) \quad \text{with} \quad K_2 = \gamma \cdot K_1^{-2}\,,$$

$$1/\varepsilon_{k+1} \geqq 1/(\varepsilon_k(1 - K_2\varepsilon_k)) = (1/\varepsilon_k)\sum_{i=0}^{\infty}(K_2\varepsilon_k)^i\,.$$

An estimation of the sum by its first two summands yields

$$1/\varepsilon_{k+1} \geqq (1 + K_2\varepsilon_k)/\varepsilon_k = K_2 + 1/\varepsilon_k$$

such that

$$1/\varepsilon_k = k \cdot K_2 + 1/\varepsilon_0, \quad \text{hence} \quad \varepsilon_k \leqq 1/k \cdot K_2\,.$$

Thus

$$\varepsilon_k = f^* - f(x^k) \leqq a_1/k$$

with

$$a_1 = 1/K_2 = (1/\gamma)\,((MN) \cdot \|\hat{x} - x^k\|/(-g(\hat{x}, t^k) - d_0))^2\,.$$

Since $f^* - f(x^k) \geqq \gamma\,\|x^k - x^*\|^2$ it follows

$$\|x^k - x^*\| \leqq (1/\gamma)\,(f^* - f(x^k))^{1/2} \leqq (\gamma \cdot k \cdot K_2)^{-1/2}\,,$$

respectively,

$$\|x^k - x^*\| \leqq a_2/\sqrt{k}$$

with

$$a_2 = (1/\gamma)\ (MN) \cdot \|\hat{x} - x^k\|)/(-g(\hat{x}, t^k) - d_0)\ . \quad \blacksquare$$

Remark 3.6.: This convergence order is the usual order which is possible for linear approximation techniques in cutting-plane methods. To overcome this slow convergence it is possible to use a hybride technique (cf. HUTH [11]).

4. Computational Realization of the Algorithm

Note that nothing but continuity of $g(x^k, \cdot)$ is assumed. For that reason the following Algorithm B is constructed to solve the auxiliary problem (H_k) by a sequence of discretized problems. That is reasonable with reference to the modelling of semi-infinite programming problems in practice (T often represents time – or space coordinates), where especially the following cases of structures of the set T are of interest:

$$T := [a, b] \subset \boldsymbol{R}, \quad T := [a, b] \times [c, d] \subset \boldsymbol{R}^2\ ,$$

(cf. TICHATSCHKE [19], [22]).

The compact set T is replaced by a finite point set $T_{h_m} \subset T$ with the density

$$\max_{t \in T} \min_{t' \in T_{h_m}} \|t - t'\| \leqq h_m\ ,$$

whose description is easy to handle for the structure of T described above. Such problems are solvable by comparison of the function values at the grid points.

Therefore, one obtains the following concrete realization:

$(H_k^{h_m}) \qquad \max \{g(x^k, t) \mid t \in T_{h_m}\}\ .$

Assumption 3.1. is satisfied if $\lim_{m \to \infty} h_m = 0$ is assumed. By modification of the steps S2 and S3 of Algorithm A it is possible to construct the following implementable algorithm:

Algorithm B:

Choose $h_1, \bar{h} > 0$, $\varepsilon_1 > 0$; $\varepsilon \in (0, 1)$ and replace in Algorithm A the steps S2 and S3 by the following ones:

S2′: *Compute t^{h_m} which exactly solves the auxiliary problem* ($H_k^{h_m}$).

S3a′: *If $g(x^k, t^{h_m}) \leqq \varepsilon_1$ go to* S3b′,
otherwise put $t^k := t^{h_m}$ and go to S4.

S3b′: *If $h_m \leqq \bar{h}$ stop, x^k is an approximate solution of the problem* (2.1),
otherwise return to S2 *with $h_m := h_m \cdot \varepsilon$.*

An analysis of the proof of convergence shows that it is possible to generate the cut from any violated constraint, but for the optimality test it is theoretical necessary to compute the most violated constraint by max $\{g(x^k, t) \mid t \in T\}$. However, this leaves open the question of generating the cut from the most violated constraint in the numerical point of view.

To overcome this fact in practice, the grid is refined if the value of the most violated constraint at the point x^k – computed at problem (H_k^{hm})–becomes sufficiently small. This approach is very time consuming if the set $T \subset \boldsymbol{R}^q$ and $q > 1$.

If $g(x^k, \cdot)$ satisfies stronger assumptions at t–like concavity or differentiability–it is possible to use more efficient iterative methods, which reflect the originality of the given problems to solve the auxiliary problems (H_k)–see column generation methods, penalty methods (cf. WOLFE [25], TICHATSCHKE [19]).

Furthermore, under certain assumptions it can be shown that Algorithm B is "stable" with respect to solutions of the perturbed subproblems (P_k), i.e. because the solutions of the auxiliary problems (H_k^{hm}) are only approximately computed, so only a perturbed cut $\tilde{s}_k$ in (P_{k-1}) can be added. Hence, in the perturbed problem $(\tilde{\mathrm{P}}_k)$ only a perturbed solution $\tilde{x}^k$ can be computed.

In the following it is proved that this solution is LIPSCHITZ-continuous with respect to the unperturbed solution x^k. To this end, let (P_{k-1}) denote the unperturbed problem with the feasible region

$$Z^{k-1} = X \cap \bigcap_{t \in T^{k-1}} \{x \mid s(x^k, t; x) \leqq 0\}$$

and let $(\tilde{\mathrm{P}}_{k-1})$ denote the perturbed problem with the feasible region

$$\tilde{Z}^{k-1} = X \cap \bigcap_{t' \in T_{h_m}} \{x \mid \tilde{s}(x^k, t'; x) \leqq 0\}$$

which arises, if the cuts are calculated only over the grid-set T_{h_m} with the density h_m.

It is assumed further that in problem (P_{k-1}) and $(\tilde{\mathrm{P}}_{k-1})$ q cuts are active. D denotes the coefficient matrix of the active constraints at the point x^k for problem (P_{k-1}) and d denotes the corresponding right hand side vector of these constraints. Respectively, $\tilde{D}$ and $\tilde{d}$ are analogously defined for the perturbed problem $(\tilde{\mathrm{P}}_{k-1})$. Then the following semi norm is considered:

$$\|D - \tilde{D}\| = \max_{1 \leqq j \leqq q} \|\nabla_x g(x^k, t_j) - \nabla_x \tilde{g}(x^k, t_j')\| = \|\nabla_x g(x^k, t_0) - \nabla_x \tilde{g}(x^k, t_0')\| ,$$
$$\|d - \tilde{d}\| = \max_{1 \leqq j \leqq q} \|\nabla_x g(x^k, t_j) - \nabla_x \tilde{g}(x^k, t_j')\| \, x^k + g(x^k, t_j) - \tilde{g}(x^k, t_j')\| .$$

Assumption 2.1. guarantees the LIPSCHITZ-continuity of $g(x, \cdot)$ and $\nabla_x g(x, \cdot)$ for all $x \in X$, where L' and L'' denote the corresponding LIPSCHITZ-constants. Then the following estimates are true:

$$\|D - \tilde{D}\| \leqq L'' \cdot \|t - t'\| \leqq L'' h_m ,$$
$$\|d - \tilde{d}\| \leqq L'' \cdot h_m \|x^k\| + L' h_m .$$

Together with $L = \max(L', L'')$ and

$$\varepsilon = \max(\|D - \tilde{D}\|, \|d - \tilde{d}\|) = L \cdot h_m (\|x^k\| + 1)$$

it is possible to use a result of DANIEL [3]:

Theorem 4.1.: *Suppose Assumption* 2.1. *and* $\mathrm{rg}(D) = \mathrm{rg}(\tilde{D})$ *are fulfilled then there exists a constant* $c > 0$ *such that*

$$\|x^k - \tilde{x}^k\| \leqq c \cdot L \cdot h_m (\|x^k\| + 1) ,$$

i.e., if the grid constant h_m *tends to zero then the perturbed solution* $\tilde{x}^k$ *converges to the exact solution* x^k *of problem* (P_{k-1}).

Remark 4.2.: The Assumption $\mathrm{rg}(D) = \mathrm{rg}(\tilde{D})$ is very stringent and in practice not realistic. $\mathrm{rg}(D) = \mathrm{rg}(\tilde{D})$ means that the grid density must be sufficiently small to ensure fulrank of $\tilde{D}$. Furthermore, the rank of D is in any case unknown. From the numerical point of view the high density of the grid yields to the ill-condition of the coefficient matrix in cutting plane methods. However, Theorem 4.1. can only be a theoretical result.

5. Computational Experiments

The present Algorithm B is implemented in the programming language FORTRAN 4000 in the operating system FOBS on a computer KRS 4200. The program has a modular structure and the subroutines for solving the problems (P_k) and (H_k) are easy to replace by other iterative methods. To demonstrate the behaviour of the Algorithm B the following examples are considered.

Example 1:

$$f(x) = (x_1 - 2)^2 + (x_2 - 0.2)^2 ;$$
$$g(x, t) = (5 \sin \pi \sqrt{t}/(1 + t^2))\, x_1 - x_2 \leqq 0 ;$$
$$X = \{x \in \boldsymbol{R}^2 \mid -1 \leqq x_1 \leqq 1,\ 0 \leqq x_2 \leqq 0.2\} ;$$
$$T = \{t \in \boldsymbol{R} \mid 0 \leqq t \leqq 8\} .$$

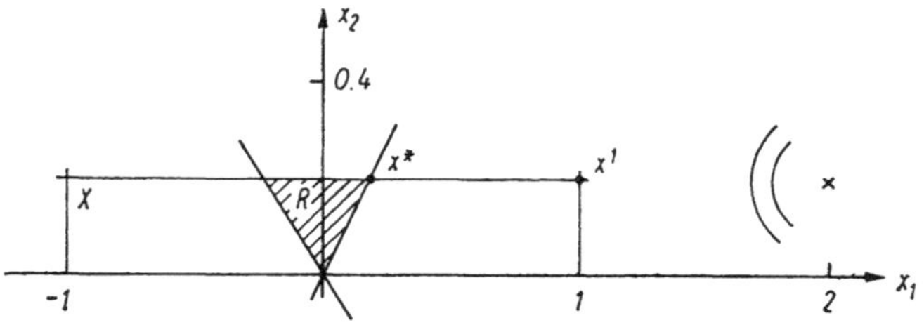

Fig. 5.1.

The tables describe the iteration sequence, where v denotes the value of the approximately determined most violated constraint

$$v = \max\{g(x^k, t) \mid t \in T_{h_m}\}$$

and r denotes the number of the constraints considered in (P_k).

Table 1: ($\bar{h}=0.01$)

k	x	$f(x)$	h	t	v	r
1	1.0	1.0	2.0	6.0	−0.066563	4
	0.2					
			1.0	6.0	−0.066563	
			0.5	0.5	2.98277	
2	0.062854	3.75253		0.5	0.000003	3
	0.200048					
			0.25	0.25	0.095737	
3	0.042487	3.83186		0.25	0.000001	3
	0.199938					
			0.125	0.25	0.000001	
			0.0625	0.1875	0.000755	
4	0.042327	3.83248		0.1875	0.000002	3
	0.199937					
			0.03125	0.21875	0.000997	
5	0.042114	3.83332		0.21875	0.0000002	3
	0.199918					
			0.015625	0.21875	0.000002	
			0.007812	0.210938	0.000033	
6	0.042111	3.83333		0.210938	0.000002	3
	0.199939					

$x^*=(0.042122,\ 0.2)$
$f(x^*)=3.83333.$

Example 2:

$$f(x)=(x_1-2)^2+(x_2-0.2)^2\ ;$$

$$g(x,t)=(5\sin\pi\sqrt{t})/(1+t^2))\,x_1^2-x_2\leqq 0\ ;$$

$$X=(x\in\boldsymbol{R}^2\mid -1\leqq x_1\leqq 1,\ 0\leqq x_2\leqq 0.2\}\ ;$$

$$T=\{t\in\boldsymbol{R}\mid 0\leqq t\leqq 8\}\ .$$

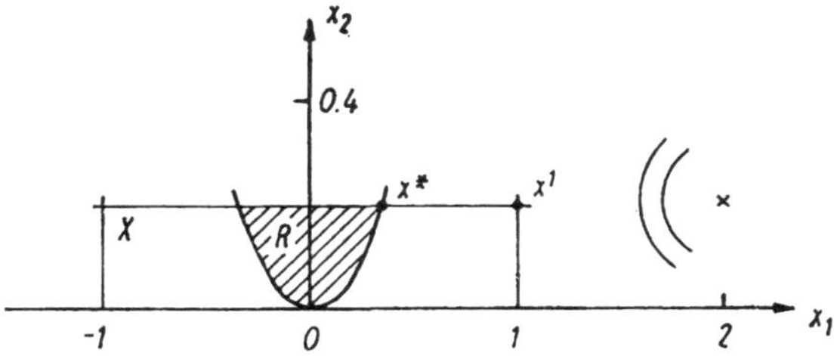

Fig. 5.2.

Table 2: ($\bar{h}=0.01$)

k	x	$f(x)$	h	t	v	r
1	1.0 0.2	1.0	2.0	6.0	−0.066563	4
			1.0	6.0	−0.066563	
			0.5	0.5	2.98277	
2	0.531420 0.200009	2.15673		0.5	0.69883	3
3	0.324862 0.200096	2.80609		0.5	0.135798	3
4	0.259235 0.200184	3.03026		0.5	0.013707	3
5	0.250953 0.200224	3.05917		0.5	0.000219	3
6	0.250765 0.200143	3.05983		0.5	−0.000001	3
			0.25	0.25	0.095777	
7	0.210100 0.199944	3.20374		0.25	0.007783	4
8	0.206326 0.200264	3.21727		0.25	0.000066	3
9	0.206240 0.200163	3.21757		0.25	0.000002	3
			0.125	0.25	0.000002	
			0.0625	0.1875	0.000756	
10	0.205696 0.199957	3.21953		0.1875	0.000003	4
			0.03125	0.21875	0.000997	
11	0.205337 0.200155	3.22083		0.21875	0.0	3
			0.015625	0.21875	0.0	
			0.007812	0.21093	0.000032	
12	0.205143 0.199808	3.22151		0.210938	−0.000004	3

$x^* = (0.20523,\ 0.2)$
$f(x^*) = 3.2212$

Example 3:

$f(x) = x_1^2 + x_2^2$;

$g(x, t) = \left((x_1-2)^2 + (x_2-2)^2 - 4\right) \cdot t_1 + (x_1^2 + x_2^2 - 4) \cdot t_2 \leqq 0$;

$X = \{x \in \boldsymbol{R}^2 \mid 0 \leqq x_i \leqq 2,\ i=1, 2\}$;

$T = (t \in \boldsymbol{R}^2 \mid 0 \leqq t_j \leqq 1,\ j=1, 2)$.

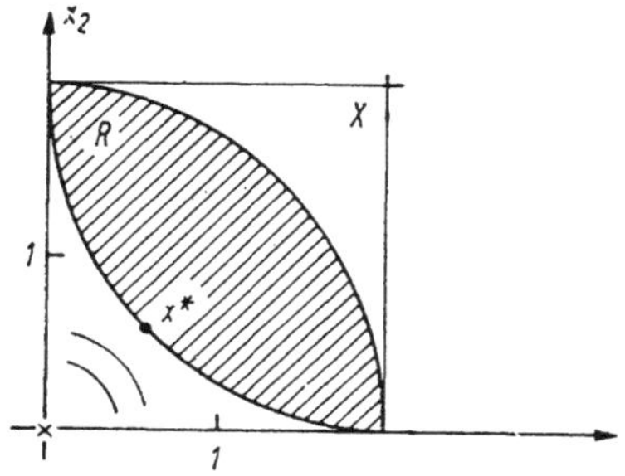

Fig. 5.3.

Table 3: ($\bar{h}=0.1$)

k	x	$f(x)$	h	t	v	r
1	0.0 0.0	0.0	0.5	1.0 0.0	4.0	4
2	0.5 0.5	0.5		1.0 0.0	0.5	3
3	0.583333 0.583333	0.680554		1.0 0.0	0.013898	2
4	0.585784 0.585784	0.686285		1.0 0.0	0.000011	2
5	0.585786 0.585786	0.686290		1.0 0.0	0.0	2
			0.25	1.0 0.0	0.0	
			0.125	1.0 0.0	0.0	
			0.0625	1.0 0.0	0.0	

$x^* = (0.585786,\ 0.585786)$
$f(x^*) = 0.686290$

As can be seen the grid T_{hm} is refined adaptively and the auxiliary problem (H_k^{hm}) is solved once more as soon as $v < \varepsilon_1 = 10\mathrm{E}-5$. The iteration is stopped only if the bound ε_1 and the given smallest grid width $\bar{h} = 0.01$ are attained. Because in the test examples the linear constraints of the set X are integrated in the dropping rules, the subproblems have the smallest possible dimension (see row r). The values of the bounds to test the activity of the constraints and to compute the approximate solution of (P_k) have to tune to another–that on the one hand each active constraint found and on the other hand the number of considered constraints in (P_k) does not get higher.

It is well known that cutting-plane methods can be instable if a great number of cuts is necessary to describe the geometry of the boundary of the feasible region. In this case the matrix of coefficients of the problem (P_k) is ill-conditioned. But it is unnecessary to compute the solution of (P_k) with the highest possible precision, because a correction of the solution is effected in the first place by grid refining, i.e. by reduction of the discretization error. For that reason the stopping bound ε_1 is not to choose too small.

Till now all examples show that Algorithm B computes stably. Its efficiency is shown in the first place by the small number of considered constraints in (P_k).

References

[1] Abadie, J. (ed.): Nonlinear programming. North-Holland Publ. Comp., Amsterdam 1967.

[2] Blankenship, J. W.; J. E. Falk: Infinitely constrained optimization problems. JOTA, 19 (1976) 261–281.

[3] Daniel, J. W.: Stability of the solution of definite quadratic programs. Mathem. Programming, 5 (1973) 1, 41–53.

[4] Eaves, B. C.; W. I. Zangwill: Generalized cutting plane algorithms. SIAM J. on Control, 9 (1971) 529–542.

[5] Grossmann, C.; A. A. Kaplan: Strafmethoden und modifizierte Lagrangefunktionen in der nichtlinearen Optimierung. Teubner-Texte, Leipzig 1979.

[6] Gonzaga, C.; E. Polak: On constraint dropping schemes and optimality functions for a class of outer approximation algorithms. SIAM J. on CONTROL, 17 (1977) 477–497.

[7] Hadley, G.: Nichtlineare und dynamische Programmierung. Verlag „Die Wirtschaft" Berlin 1969.

[8] Hettich, R. (ed.): Semi-infinite programming. Proc. of a work-shop, Bad Honnef, 1978. *In:* Lecture Notes in Control and Information Science 15, 1979.

[9] Hettich, R.; P. Zenke: Numerische Methoden der Approximation und semi-infinite Optimierung. Teubner, Stuttgart 1982.

[10] Huth, M.: Überlinear konvergente Verfahren zur Lösung nichtlinearer semi-infiniter Optimierungsaufgaben. Diss. A, PH Halle 1987.

[11] Kapur, K. C.: On cutoff optimization methods in infinite dimensional spaces and applications. JOTA, 12 (1973) 15–31.

[12] Kelley, J. E.: The cutting plane method for solving convex programs. J. of Society for Industrial and Appl. Mathematics, 8 (1960) 703–712.

[13] Koch, W.: Kleinrechnersoftware zur Behandlung nichtlinearer Optimierungsaufgaben. Dok., TU Dresden, 1984.

[14] Levitin, E. S.; D. T. Poljak: Restringierte Minimierungsmethoden. Num. Math. und mathem. Physik, 6 (1966) 1–50, (russ.).

[15] Oettli, W.: Solving optimization problems with many constraints by a sequence of subproblems containing two constraints. Math. Nachrichten, 71 (1976) 143–145.

[16] Polak, E.; A. L. Tits: A recursive quadratical convergent algorithm for semi-infinite problems. Internat. Symp. on Semi-infinite Programming, Austin 1981.

[17] Schwettlick, H.: Numerische Lösung nichtlinearer Gleichungen. Verlag der Wissenschaften, Berlin 1979.

[18] Tichatschke: Semi-infinite optimization. *In:* Mathematical Methods in Operation Research, Bulg. Acad. of Science, Sofia 1987, 155–171.

19] TICHATSCHKE, R.: Lineare semi-infinite Optimierungsaufgaben und ihre Anwendungen in der Approximationstheorie. Wissenschaftliche Schriftenreihe, TH Karl-Marx-Stadt, Heft 4, 1981.

[20] TICHATSCHKE, R.; TH. LOHSE: Eine verallgemeinerte Schnittmethode für konvexe semi-infinite Optimierungsaufgaben. Wiss. Zeitschrift TH Karl-Marx-Stadt, **24** (1982) 3, 332–338.

[21] TICHATSCHKE, R.: The numerical treatment of semi-infinite problems. Banach-Center-Publ., 6 (1984), Warszawa.

[22] TICHATSCHKE, R.; H. KIRSTEN: The efficiency of a method of feasible directions for solving variational inequalities. *In:* Lecture Notes in Economics and Mathem. Systems, **294** (1986) 379–400, Springer-Verlag.

[23] TOPKIS, D. M.: Cutting-plane methods without nested constraint sets. Operations Research, **18** (1970), 404–413.

[24] TOPKIS, D. M.: A cutting-plane algorithm with linear and geometric rate of convergence. JOTA, **36** (1982) 1–22.

[25] WOLFE, P.: Methods of nonlinear programming. *In:* [1], 99–130.

Received March 1987, revised January 1988

R. Tichatschke
Pädagogische Hochschule
"Liselotte Herrmann" Güstrow
Goldberger Str. 12
Güstrow
2600

V. Nebeling
Pädagogische Hochschule
"N. K. Krupskaja" Halle
Kröllwitzer Str. 44
Halle (Saale)
DDR - 4020

optimization **19** (1988) 6, 818

Book Review

WALTER, W. (ed.): **General Inequalities 5.** Birkhäuser Verlag Basel, 1987, 502 S., DM 114,–, ISBN 3-76543-1799-X

This volume reflects, that both classical and modern inequalities continue to play a significant role in nearly all areas of mathematics and also that they continue to influence developments in mathematics.

The first article contains a report on DE BRANGE'S proof of the BIEBERBACH conjecture. Then the articles are divided into groups. The first group contains inequalities for sums, series and integrals and we find convolution inequalities, a HARDY-LITTLEWOOD type integral inequality, inequalities in connection with recurrence relations of dynamic programming type, quadratic inequalities, inequalities for geometric means and estimations of singular integrals.

In the next group the articles deal with inequalities in analysis and approximation (weighted integral inequalities, estimates for convolutions of kernel functions, inequalities for some special functions and their zeros, harmonic L^1 – approximation, properties of (s, t) – convex functions and of moments of convex functions).

The third group is dedicated to inequalities of functional analysis (summability, FOURIER and operator inequalities, estimates for ultraproducts in BANACH spaces).

The next papers contain functional equations and inequalities (stability, iterative inequalities, CHOLEWA inequality, assumptions under which a special difference equation has exactly one convex solution). Then a group of inequalities for differential operators follows. We find discrete GRONWALL type inequalities, bounds of eigenvalues (several papers) and isoperimetric inequalities, generalizations of the HARDY-LITTLEWOOD-PÓLYA inequality, remarks on K in $\|y'\|^2 < K\|y\| \cdot \|y''\|$, an existence theorem for ordinary differential equations in ordered BANACH spaces and at least SOBOLEW type inequalities.

Finally inequalities from other fields are treated (maximum entropy principle, generalized HÖLDER inequalities with connections to optimization problems, extension of SCHUR'S majorization theorem, a HAHN-BANACH-theorem for abelian semigroups, symmetric distribution functions).

A variety of interesting problems is to be found in the problems and remarks section (inequalities for means, SCHUR convexity, YOUNG'S inequality, packings of congruent balls in the unit ball).

The index at the end of the book together with some artistic sketches (by JOY RUSSEL) is a nice and useful completion. I would like to recommend this volume to each mathematician.

A. GÖPFERT

optimization 19 (1988) 6, 819–828

A Global Convergence Theorem for Aggregation Algorithms

L. García

Inst. de Cibernética Matemática y Física, Academia de Ciencias, Habana, Cuba

Summary: In this paper conditions are given to assure global convergence for a broad class of aggregation algorithms of linear programming. For a more restrictive class properties of the accumulation points are also given, even if global convergence conditions are not satisfied.

AMS 1980 Subject Classification:

Primary: 49 B 36; Secondary: 90 C 06.

Key words: Iterative aggregation, linear programming, large-scale, global convergence.

1. Introduction

For L. P. problems, aggregation has often been used to substitute the original large problem by a smaller one. From this point of view the interest has been focussed on finding bounds for the error, using such an approximate model [see, e.g., Zipkin [12] and Mendelssohn [7]).

Another point of view has been to elaborating iterative aggregation algorithms, i.e., iterative processes of adjustment of the aggregation model in order to obtain approximate solutions of the original problem as good as desired. In this sense convergence properties are of a great interest.

Vakhutinsky and Dudkin [10] have proposed an iterative aggregation algorithm for the general L. P. problem, for which the optimality of stationary points could be proved. Some years later Vakhutinsky et al. [11] have given a modification of these processes for which local convergence has been proved under the hypothesis of the uniqueness of the optimum as well for the primal problem as for its dual.

All these algorithms and also the soviet literature on this subject, make use of modified Lagrange functions and the aggregate problems will be defined as quadratic ones with non-negative variables.

Another approach is followed by Martínez [6], and he proves global convergence under the assumption that the feasible region has a non-empty interior. Similar ideas are posteriorly used by Garcia [2, 3] and by Mendelssohn [8].

In this paper we face the problem of convergence of iterative aggregation algorithms for an arbitrary L. P. problem from a general point of view, and a global convergence theorem is obtained.

2. Aggregate Problem

The original L. P. problem is

$$\boldsymbol{P}: \qquad \max \{cx \mid x \in \mathbb{R}^n,\ x \geqq 0,\ Ax \leqq b\},$$

where A is an $m \times n$-matrix, c is an n-dimensional row-vector, and b is an m-dimensional column-vector.

We suppose $\boldsymbol{P}$ to have a finite optimum and we denote by Z^* the optimal value of $\boldsymbol{P}$, and by F (respectively F^*) the set of all feasible (respectively optimal) solutions of $\boldsymbol{P}$.

Let $J := \{1, \ldots, n\}$ be the column-index set.

For a given vector $y^0 \in \mathbb{R}^n$ and a partition $\mathfrak{Q} := (J_s)_{s=1,\ldots,r}$ of J – i.e. $\bigcup_{s=1}^{r} J_s = J$ and $J_s \cap J_k = \emptyset$ for all $1 \leqq k, s \leqq r, k \neq s$ –, we define the column aggregate problem $\tilde{\boldsymbol{P}}(y^0, \mathfrak{Q})$ as follows:

$$\tilde{\boldsymbol{P}}(y^0, \mathfrak{Q}): \quad \geqq \max \left\{ \sum_{s=1}^{r} \bar{c}_s t_s \mid t_s \geqq 0,\ s = 1, \ldots, r,\ \sum_{s=1}^{r} \bar{a}^s t_s \leqq b \right\}.$$

where the coefficients $\bar{c}_s$, $\bar{a}^s$ of the aggregate variable t_s are defined by

$$\bar{c}_s := \sum_{j \in J_s} c_j y_j^0, \quad \bar{a}^s := \sum_{j \in J_s} a^j y_j^0, \quad \text{where } a^j \text{ denotes the } j\text{-th}$$

column of matrix A.

The *aggregation vector* y^0 is always supposed to be non-negative. We shall denote by $F(y^0, \mathfrak{Q})$ (respectively $F^*(y^0, \mathfrak{Q})$) the set of feasible (respectively optimal) solutions of $\tilde{\boldsymbol{P}}(y^0, \mathfrak{Q})$. We denote the dual problem of $\tilde{\boldsymbol{P}}(y^0, \mathfrak{Q})$ by $\tilde{\boldsymbol{P}}'(y^0, \mathfrak{Q})$.

Given an r-vector t, we call the *disaggregation* of t *by means of* y^0 *and* $\mathfrak{Q}$, the n-vector x defined by

$$x^s := y^{0s} t_s \quad \text{for all} \quad s = 1, \ldots, r. \tag{1}$$

where x^s, y^{0s} denote the components of x and y^0 in J_s. If t is a feasible solution of $\tilde{\boldsymbol{P}}(y^0, \mathfrak{Q})$ then its disaggregation x is a feasible solution of the original problem. Moreover,

$$\bar{c}t = cx, \tag{2}$$

i.e., both solutions have the same evaluation in their correspondent problems.

Given a partition $\mathfrak{Q}$, we shall say that a non-negative vector $y^0 \in \mathbb{R}^n$ is a feasible aggregation vector for $\boldsymbol{P}$ with respect to $\mathfrak{Q}$, if $\tilde{\boldsymbol{P}}(y^0, \mathfrak{Q})$ has a feasible solution.

If $g(y^0, \mathfrak{Q})$ denotes the optimal value of $\tilde{\boldsymbol{P}}(y^0, \mathfrak{Q})$ it holds from (2) that if y^0 is a feasible aggregation vector, then we have

$$-\infty < g(y^0, \mathfrak{Q}) \leqq Z^*. \tag{3}$$

So we could equivalently say that y^0 is a feasible aggregation vector if $\tilde{\boldsymbol{P}}(y^0, \mathfrak{Q})$ has an optimal solution.

If the equality holds in (3), i.e. the optimal values of $\boldsymbol{P}$ and $\tilde{\boldsymbol{P}}(y_0, \mathcal{Q})$ coincide, we shall say that y^0 is an optimal aggregation vector.

Let be $S(\boldsymbol{P}, \mathcal{Q})$, respectively $S^*(\boldsymbol{P}, \mathcal{Q})$, the set of all feasible, respectively optimal, aggregation vectors for $\boldsymbol{P}$ with respect to $\mathcal{Q}$. Note that, in particular, $F \subset S(\boldsymbol{P}, \mathcal{Q})$ and $F^* \subset S^*(\boldsymbol{P}, \mathcal{Q})$.

3. A Global Convergence Theorem

We suppose a partition $\mathcal{Q}$ fixed. Let us consider a general aggregation algorithm $\mathcal{A}$ (see [5, p. 121]) defined on $S(\boldsymbol{P}, \mathcal{Q})$, such that it can be described as the composition of two point-to-set mappings, i.e. $\mathcal{A} = \mathcal{T} o \mathcal{G}$, where

– $\mathcal{G}(y)$ is defined for each $y \in S(\boldsymbol{P}, \mathcal{Q})$ as the set of all vectors (x, u) such that x is the disaggregation of an optimal solution, t, of $\tilde{\boldsymbol{P}}(y, \mathcal{Q})$, and u is an optimal solution of $\tilde{\boldsymbol{P}}$, $(y, \mathcal{Q})$.

– point-to-set mapping $\mathcal{T}$ is defined for all $(x, u) \in \operatorname{Im} \mathcal{G}$, and satisfies:

a) $\mathcal{T}(x, u) \subset S(\boldsymbol{P}, \mathcal{Q})$ for all $(x, u) \in \operatorname{Im} \mathcal{G}$.

b) $g(z, \mathcal{Q}) \geqq g(x, \mathcal{Q})$ for all $z \in \mathcal{T}(x, u)$.

$\mathcal{G}(y)$ gives us information about the aggregation vector y: the best feasible solution x of $\boldsymbol{P}$ which can be obtained by means of y and some elements to evaluate how good is x, namely prices $u = (u_i)_{i=1,\ldots,m}$. Then considering this information, we can correct y in a certain way (algorithm $\mathcal{T}$). Thus, algorithm $\mathcal{A}$ works in two phases at each iteration: first, it looks for information about y by solving $\tilde{\boldsymbol{P}}(y, \mathcal{Q})$, and then it corrects y accordingly, if necessary.

Let $(y^i\}$ be a sequence constructed by algorithm $\mathcal{A}$ starting from $y^0 \in S(\boldsymbol{P}, \mathcal{Q})$, that is

$$y^i \in \mathcal{T}(x^i, u^i) \quad \text{for some} \quad (x^i, u^i) \in \mathcal{G}(y^{i-1}) .$$

From the hypotheses above it holds

– $y^i \in S(\boldsymbol{P}, \mathcal{Q})$ for all $i \in \mathbb{N}$.

– $g(y^i, \mathcal{Q})$ is a monotonically non-decreasing sequence, since

$$g(y^{i+1}, \mathcal{Q}) \geqq g(x^i, \mathcal{Q}) = g(y^i, \mathcal{Q}) .$$

In other words, $\mathcal{A}$ is a class of aggregation algorithms which construct at each iteration a new feasible aggregation vector, as least as better as the previous one.

Obviously, point-to-set mapping $\mathcal{T}$ characterizes each particular case of the general algorithm $\mathcal{A}$, and it determines its convergence. However, in relation to the study of the convergence of algorithm $\mathcal{A}$, it is to note that the point-to-set mapping $\mathcal{G}$ is not closed. For example, we may have

$$y^i \to y^*, \quad (x^i, u^i) \in \mathcal{G}(y^i) \quad \text{for all} \quad i \in \boldsymbol{N},$$

and $(x^i, u^i) \to (x^*, u^*)$, while for some $s \in \{1, \dots, r\}$ it holds

$$y^{is} := (y^i_j)_{j \in J_s} \to 0$$

and

$$x^{is} := (x^i_j)_{j \in J_s} \to x^{*s} \neq 0 .$$

So x^* would not belong to $\mathcal{G}(y^*)$.

Another difficulty arises from the non-necessary convexity of $S(\boldsymbol{P}, \mathfrak{Q})$, and $S^*(\boldsymbol{P}, \mathfrak{Q})$ that we have proved in [1].

In the following we shall suppose F to be bounded.

Lemma 1: *Let $\{y^i\} \subset S(\boldsymbol{P}, \mathfrak{Q})$, and for each $i \in \mathbb{N}$ let t^i, u^i be optimal solutions of $\tilde{\boldsymbol{P}}(y^i, \mathfrak{Q})$ and its dual respectively, and let x^i be the disaggregation of t^i. If for some subsequence $K \in \mathbb{N}$ it holds $y^i \xrightarrow[i \in K]{} y^*$, $u^i \xrightarrow[i \in K]{} u^*$, then any accumulation point x^*, of $\{x^i\}_{i \in K}$ satisfies $cx^* = g(x^*, \mathfrak{Q}) \geqq g(y^*, \mathfrak{Q})$.*

Proof: Since F is compact, we can suppose $(x^i\}_{i \in K}$ to be convergent without any loss of generality.

On the one hand, since $x^* \in F$, it holds

$$g(x^*, \mathfrak{Q}) \geqq cx^* . \tag{4}$$

On the other hand, for each $i \in \mathbb{N}$, u^i satisfies

$$u^i \geqq 0 , \quad u^i b = cx^i \tag{5a}$$

$$u^i(A^s x^{is}) \geqq c^s x^{is} \quad \text{for all} \quad s = 1, \dots, r. \tag{5b}$$

Therefore, taking limit in (5) we obtain that u^* is a feasible solution of $\tilde{\boldsymbol{P}}'(x^*, \mathfrak{Q})$ and $cx^* = u^*b$. Thus according to (4), cx^* is the optimal value of $\tilde{\boldsymbol{P}}(x^*, \mathfrak{Q})$, i.e. $g(x^*, \mathfrak{Q}) = cx^*$. Analogously, u^* is also a feasible solution of $\tilde{\boldsymbol{P}}'(y^*, \mathfrak{Q})$, so $g(y^*, \mathfrak{Q}) \leqq$ $\leqq u^*b$, which completes the proof. ■

Since we have $cx^i = g(y^i, \mathfrak{Q})$ by construction, then it holds by this lemma

$$g(y^*, \mathfrak{Q}) \leqq cx^* \leqq \lim_{i \in K} g(y^i, \mathfrak{Q}) . \tag{6}$$

Imposing some additional conditions to the point-to-set mapping $\mathfrak{T}$, (6) can be improved as follows:

Theorem 1: *Let point-to-set mapping $\mathfrak{T}$ be such that*

i) $\mathfrak{T} \circ \mathcal{G}(y) \subset F$ for all $y \in S(\boldsymbol{P}, \mathfrak{Q})$.

ii) $c\bar{y} \geqq g(y, \mathfrak{Q})$ for all $\bar{y} \in \mathfrak{T} \circ \mathcal{G}(y)$.

Then if $y^i \xrightarrow[i \in K]{} y^*$ and $(x^i, u^i) \xrightarrow[i \in K]{} (x^*, u^*)$, it holds

a) $cx^* = g(x^*, \mathfrak{Q})$.

b) $cx^* = cy^* = g(y^*, \mathfrak{Q}) = \lim_{i \in K} g(y^i, \mathfrak{Q})$.

Proof: (a) follows from Lemma 1.

In relation to (b), conditions (i) and (ii) imply

$$cy^i \leqq g(y^i, \mathfrak{Q}) = cx^i \leqq cy^{i+i} .$$

Therefore, for $i \in K$ we obtain in the limit

$$cy^* = \lim_{i \in K} g(y^i, \mathcal{Q}) = cx^* . \tag{7}$$

But, since $y^* \in F$, then

$$cy^* \leqq g(y^*, \mathcal{Q}) ; \tag{8}$$

thus, from (6–8), (b) holds. ■

Theorem 2 (*Global convergence*): *Under the hypothesis of Theorem* 1, *and if the following conditions hold:*
iii) $\mathcal{T}$ *is closed.*
iv) *If* $y^i \notin S^*(\boldsymbol{P}, \mathcal{Q})$ *then* $cy > g(y^i, \mathcal{Q})$ *for all* $y \in T \circ \mathcal{G}(y^i)$.
Then x^* *and* y^* *are optimal solutions of* $\boldsymbol{P}$.

Proof: Since $\{y^i\} \subset F$ and F is bounded, we can take without loss of generality $y^{i+1} \xrightarrow[i \in K]{} \hat{y}$.
Then, since $(x^i, u^i) \xrightarrow[i \in K]{} (x^*, u^*)$ and $y^{i+1} \in \mathcal{T}(x^i, u^i)$ for all $i \in K$, we have $\hat{y} \in \mathcal{T}(x^*, u^*)$ since $\mathcal{T}$ is closed. But from hypothesis (ii) the sequence $\{cy^i\}_{i \in \mathbb{N}}$ is monotonically non-decreasing so $cy^* = c\hat{y}$; and from Theorem 1

$$cx^* = g(x^*, \mathcal{Q}) = cy^* . \tag{9}$$

Therefore

$$(x^*, u^*) \in \mathcal{G}(x^*) \quad \text{and} \quad c\hat{y} = g(x^*, \mathcal{Q}) , \quad \text{where} \quad \hat{y} \in \mathcal{T} \circ \mathcal{G}(x^*) .$$

Then, according to (iv) x^* is an optimal aggregation vector. But, since x^* and y^* are feasible solutions of $\boldsymbol{P}$ it follows from (9) that x^* is an optimal solution of $\boldsymbol{P}$ and so do y^*. ■

4. A Particular Class of Aggregation Algorithms

Now we consider a particular case of algorithm $\mathcal{A}$ above, where the improvement of the aggregation vector is made by solving an auxiliary L. P. problem whose coefficients depend on the parameters (x, u).

Briefly, we suppose that for each (x, u) an L. P. problem can be defined as follows:

$$\bar{\boldsymbol{P}}(x, u) \colon \max \{\bar{c}(x, u) \mid z \in \mathbb{R}^{n_1}, z \geqq 0, \bar{A}(x, u)\, z \leqq \bar{b}(x, u)\} ,$$

where $\bar{A}(x, u)$ is an $m_1 \times n_1$ matrix and all the coefficients of $\bar{\boldsymbol{P}}(x, u)$ depend continuously on the parameters (x, u). Moreover, $\bar{\boldsymbol{P}}(x, u)$ is such that if x is a feasible solution of $\boldsymbol{P}$ and z is an optimal solution of $\bar{\boldsymbol{P}}(x, u)$, then $\bar{c}(x, u)\, z \geqq cx$. We suppose also, that for any $(x, u) \in \mathbb{R}^n \times \mathbb{R}^m$ there exists a function $f_{(x,u)} : \mathbb{R}^{n_1} \to \mathbb{R}^n$ which satisfies:
a) $f_{(x,u)}(z)$ is a continuous function of (x, u, z).

b) For any feasible solution, z, of $\bar{\boldsymbol{P}}(x, u)$ the vector $f_{(x,u)}(z)$ is a feasible solution of $\boldsymbol{P}$, and

$$\bar{c}(x, u)\, z = c[f_{(x,u)}(z)] \;,$$

i.e. $f_{(x,u)}$ preserves the feasibility and the evaluation of any z.

The algorithm considered is then $\mathcal{A}_1 = \mathcal{T}_1 \circ \mathcal{G}$, where the point-to-set mapping $\mathcal{T}_1$ is defined as the image-set by function $f_{(x,u)}$ of the set of optimal solutions of $\bar{\boldsymbol{P}}(x, u)$.

To have a well defined algorithm $\mathcal{A}_1$, we shall suppose that $\bar{\boldsymbol{P}}(x, u)$ has a feasible solution for any $(x, u) \in \mathcal{G}(y)$, and all $y \in S(\boldsymbol{P}, \mathcal{Q})$. Since F is bounded, this means accordingly to condition (b), that $\bar{\boldsymbol{P}}(x, u)$ will also have an optimal solution. Clearly, algorithm $\mathcal{A}_1$ satisfies conditions (i) and (ii) of Theorem 1.

Before applying the results of Section 3 to this case let us prove the following

Lemma: *Let $\{y^i\} \subset S(\boldsymbol{P}, \mathcal{Q})$, and $(x^i, u^i) \in \mathcal{G}(y^i)$ for all $i \in \mathbb{N}$, and let z^i and w^i be optimal solutions of $\bar{\boldsymbol{P}}(x^i, u^i)$ and its dual respectively. Then if $(x^i, u^i) \xrightarrow[i\in K]{} (x^*, u^*)$ and $(z^i, w^i) \xrightarrow[i\in K]{} (z^*, w^*)$ it holds that z^* is an optimal solution of $\bar{\boldsymbol{P}}(x^*, u^*)$.*

Proof: Since all coefficients of $\bar{\boldsymbol{P}}(x, u)$ are continuous on (x, u), it follows that z^* and w^* are feasible solutions of $\bar{\boldsymbol{P}}(x^*, u^*)$ and its dual respectively; moreover, they have the same value, so they are optimal ones. ■

Remark: With the notations above, if we would have that if the sequence $\{f_{(x^i,u^i)}(z^i)\}$ is convergent then $\{z^i\}_{i\in K}$ has a convergent subsequence, then it follows from this lemma that point-to-set mapping $\mathcal{T}_1$ would be closed.

Theorem 3: (*Theorem of equilibrium*)*: Let $(y^i\}$ be a sequence constructed by algorithm $\mathcal{A}_1$ above, using $\{(x^i, u^i)\}$ and $\{(z^i, w^i)\}$. If for a subsequence $K \subset \mathbb{N}$ we have $y^i \xrightarrow[i\in K]{} y^*$, $(x^i, u^i) \xrightarrow[i\in K]{} (x^*, u^*)$ and $(z^i, w^i) \xrightarrow[i\in K]{} (z^*, w^*)$, then cx^* is the optimal value of $\tilde{\boldsymbol{P}}(x^*, \mathcal{Q})$* and *$\bar{\boldsymbol{P}}(x^*, u^*)$.*

Proof: From Theorem 1 we have that cx^* is the optimal value of $\tilde{\boldsymbol{P}}(x^*, \mathcal{Q})$. and

$$cx^* = cy^* \; . \tag{10}$$

But since $\{cy^i\}$ is a monotonically non-decreasing sequence, we have $cy^* = \lim_{i\in\mathbb{N}} cy^i$.

On the other hand $cy^{i+1} = \bar{c}(x^i, u^i)\, z^i$ for all $i \in \mathbb{N}$, thus

$$\bar{c}(x^*, u^*)\, z^* = \lim_{i\in K} cy^{i+1} = cy^* \; . \tag{11}$$

Then, from Lemma 2 and relations (10) and (11), it follows that cx^* is the optimal value of $\bar{\boldsymbol{P}}(x^*, u^*)$. ■

This theorem states that any accumulation point x^* of a sequence $(x^i\}$ obtained by algorithm $\mathcal{A}_1$, is a stationary one. If we think of algorithm $\mathcal{A}_1$ as an alternative action of problems $\tilde{\boldsymbol{P}}$ and $\bar{\boldsymbol{P}}$ in order to improving the starting aggregation vector y^0, then such an x^* represents a state of equilibrium which algorithm $\mathcal{A}_1$ tends to obtain.

Analogously to Theorem 2 we have in this particular case the following result:

Theorem 4: (*Global convergence*): *Under the hypothesis of Theorem 3 and condition (iv) of Theorem 2, then x^* and y^* are optimal solutions of $\boldsymbol{P}$.*

Proof: According to Theorem 3 and condition (iv) it follows inmediately that x^* is an optimal solution of $\boldsymbol{P}$.

On the other hand, y^* is a feasible solution of $\boldsymbol{P}$, and we know from Theorem 1 that $yc^* = cx^*$, so y^* is also an optimal solution. ∎

Note that we have not needed $\mathcal{T}_1$ to be closed.

5. An Economic Application

In order to illustrate Theorem 3, let us consider the following aggregation algorithm, that could also be viewed as a resource directive decomposition algorithm, (see [9], for example), where the adjustment of the resource distribution is mainly made be solving an aggregate problem.

We shall take $\boldsymbol{P}$ as before, and $\mathcal{Q} := (J_k)_{k=1,\ldots,r}$ a partition of the index-set J.

Algorithm 1

Step 1. Select $y^0 \in S(\boldsymbol{P}, \mathcal{Q})$. Set $i := 0$.

Step 2. Solve $\tilde{\boldsymbol{P}}(y^i, \mathcal{Q})$. Let x^i be the disaggregation of an optimal solution of $\tilde{\boldsymbol{P}}(y^i, \mathcal{Q})$. Set $s^i := b - Ax^i$ (the resources not used by plan x^i).

Step 3. Distribute the resources s^i among the r sectors, i.e., select r vectors h^{ik}, $k = 1, \ldots, r$, from $\mathbb{R}^m$, such that

$$h^{ik} \geqq 0 \quad k = 1, \ldots, r; \quad \sum_{k=1}^{r} h^{ik} \leqq s^i .$$

Step 4. For each sector J_k, $k = 1, \ldots, r$, find the optimal use of the resources assigned to it $-Ax^{ik} + h^{ik}-$, by solving

$$\boldsymbol{P}_k(x^i, h^{ik}) : \max \{c^k z^k \mid z^k \in \mathbb{R}^{J_k},\ z^k \geqq 0,\ A^k z^k \leqq A^k x^{ik} + h^{ik}\}$$

$k = 1, \ldots, r$.

(Note: c^k, A^k and x^{ik}, denote respectively the components of c, A, and x^i in J_k).

Let z^{ik} be an optimal solution of $\boldsymbol{P}_k(x^i, h^{ik})$, $k = 1, \ldots, r$.

Step 5. Define $y^{i+1} \in \mathbb{R}^n$ as follows:

$$y_j^{i+1} := z_j^{ik} \quad \text{for} \quad j \in J_k\,, \quad k = 1, \ldots, r\,.$$

Set $i := i + 1$, and go to Step 2.

Remark: In particular, for $h^{ik}=0$, $k=1, \ldots, r$, vector y^{i+1} defined above is simply the *optimal disaggregation* of x^i. Thus, we can consider Step 4, as a disaggregation of x^i, but in a more general way.

Then, this algorithm can be viewed as an alternative process of aggregation and disaggregation in a wider sense. In particular, with optimal disaggregation we know (see García [1]) that this process doesn't converge to an optimal solution in general; therefore, the question is what happen if we take $\sum_{k=1}^{r} h^{ik} \neq 0$, for all i. For this analysis we shall use the results of Section 4.

In effect, if we consider vectors h^{ik}, $k=1, \ldots, r$, as continuous functions of x^i, it can be easily verified that Algorithm 1 is simply a particular case of the general Algorithm $\mathcal{A}_1$ described in Section 4. Problem $\tilde{\boldsymbol{P}}(x^i, u^i)$ would be in this case

$$\max \sum_{k=1}^{r} c^k z^k$$

s. t.

$$\begin{pmatrix} A^1 & & \\ & \ddots & \\ & & A^r \end{pmatrix} \begin{pmatrix} z^1 \\ \vdots \\ z^r \end{pmatrix} \begin{matrix} \leqq \\ \vdots \\ \leqq \end{matrix} \begin{pmatrix} A^1 x^{i1} + h^{i1}(x^i) \\ \vdots \\ A^r x^{ir} + h^{ir}(x^i) \end{pmatrix}$$

$$z^k \in \mathbb{R}^{J_k} \quad z^k \geqq 0 \quad k=1, \ldots, r$$

and $f_{(x^i,u^i)}(\cdot)$ will be the permutation of $\mathbb{R}^n$ defined in Step 5 of Algorithm 1.

All the initial conditions described in Section 4 can be easily verified in this case. From Theorem 3 we obtain:

Theorem 5: *Let $\{y^i\}$ be a sequence constructed by Algorithm 1, using $\{(x^i, u^i)\}$ and $\{(z^i, w^i)\}$ in the sense of Theorem 3–$z^i := (z^i{}_j)_{j=1,\ldots,k}$ –. If for a subsequence $K \subset \mathbb{N}$ we have $y^i \xrightarrow[i \in K]{} y^*$, $(x^i, u^i) \xrightarrow[i \in K]{} (x^*, u^*)$ and $(z^i, w^i) \xrightarrow[i \in K]{} (z^*, w^*)$, then*

1. *cx^* is the optimal value of $\boldsymbol{P}(x^*, \mathfrak{Q})$, i.e., $e := (1, \ldots, 1) \in \mathbb{R}^r$ is an optimal solution of $\tilde{\boldsymbol{P}}(x^*, \mathfrak{Q})$.*
2. *For each $k=1, \ldots, r$, x^{*k} is an optimal solution of $\boldsymbol{P}_k(x^*, h^{*k})$, where $h^* = (h^{*k})_{k=1,\ldots,r}$ is an arbitrary accumulation point of the sequence $\{h^i = (h^{ik})_{k=1,\ldots,r}\}_{i \in K}$, and there exists one.*

Proof: Since $\{h^i\}_{i \in K}$ is a bounded sequence, Theorem 3 can be applied even if the right-hand side of $\tilde{\boldsymbol{P}}(x^i, \mathfrak{Q})$ is not a continuous function of x^i. ■

Remark: Since we have supposed F to be bounded, if the sets

$$F_k := \{u \in \mathbb{R}^m\} \; uA^k \geqq c^k, \; u \geqq 0\}, \quad k=1, \ldots, r$$

and the sequence $\{c^i\}_{i \in \mathrm{N}}$ are also supposed to be bounded, Theorem 5 can be stated as follows:

"Any accumulation point, x^*, of a sequence $\{x^i\}_{i \in \mathrm{N}}$ constructed by the algorithm, verifies conditions (1) and (2) above."

A vector x^* that satisfies conditions (1) and (2) of Theorem 5, with $h^i \equiv 0$, $i \in \mathbb{N}$, will be called a *partially stable* – or simply *stable* – solution with respect to $\mathfrak{Q}$.

It is easy to see that if a vector x^* is stable and satisfies conditions (2) of Theorem 5 with $h^{*k} := s^*$, $k=1, \ldots, r$, then x^* satisfies it for any $0 \leqq h^{*k} \leqq s^*$, $k=1, \ldots, r$. Such an x^* will be called a *strongly stable* solution, with respect to $\mathfrak{Q}$.

For some choices of $\{h^{ik}\}$ in Algorithm 1, it can be proved that the accumulation points, according to Theorem 5, are precisely, strongly stable solutions. This is the case, for example, if we define, for some $0 < \vartheta_0 < 1$ fixed:

$$h^{ik} := \vartheta^{ik} s^i \quad k=1, \ldots, r, \quad i \in \mathbb{N}$$

where

$$\vartheta^{ik} \in \mathbb{R}; \quad \vartheta^{ik} \geqq \vartheta_0 > 0 \quad k=1, \ldots, r; \quad \sum_{k=1}^{r} \vartheta^{ik} \leqq 1 .$$

Effectively, from condition (2) of Theorem 5 it follows that x^{*k}, is a feasible solution of $\boldsymbol{P}_k(x^*, s^*)$ for all $k=1, \ldots, r$; moreover, any feasible solution, $\bar{u}$, of $\boldsymbol{P}'_k(x^*, h^*)$ is feasible of $\boldsymbol{P}'_k(x^*, s^*)$ and satisfies $\bar{u}h^{*k}=0$. But, since $\bar{u} \geqq 0$, and $h^{*k} \geqq \vartheta_0 s^* \geqq 0$, $k=1, \ldots, r$, it follows that $\bar{u}s^*=0$ and therefore x^{*k} will be optimal solution of $\boldsymbol{P}_k(x^*, s^*)$ for each $k=1, \ldots, r$.

In the following case the conditions of equilibrium, (1) and (2), of Theorem 5 coincide with the conditions of strong stability:

Let $\mathfrak{Q} = (J_k)_{k=1,\ldots,r}$ and $(y^i)_{i \in \mathrm{N}}$ be such that for $k=2, \ldots, r$, it holds

$$J_k = \{j_k\} , \quad y^{ik} = y^i_{j_k} = 1 .$$

That is, we aggregate only columns $(a^j)_{j \in J_1}$ – in one column –, and the others columns of the original problem remain unchanged. Let us also define

$$h^{i1} \equiv s^i , \quad h^{ik} = 0 , \quad k=2, \ldots, r \quad \text{for all} \quad i .$$

Then, each x^{ik}, $k=2, \ldots, r$ is also an optimal solution of $\boldsymbol{P}_k(x^i, s^i)$ since if it is not, we get a contradiction with the fact that cx^i is the optimal value of $\tilde{\boldsymbol{P}}(y^i, \mathfrak{Q})$.

Briefly, the conditions satisfied by the accumulation points of Algorithm 1 are at least conditions of stability and at most of strong stability.

In GARCÍA [4, p. 62–71] some properties and examples about stable and strongly stable solutions are given, and, in particular it is shown that even the strong stability with respect to a given partition is not, in general, a sufficient condition of optimality. Thus Algorithm 1 is not necessarily convergent.

6. Conclusions

Theorems 1–4 give a methodological tool for the study of the convergence of the wide class of aggregation algorithms considered.

This was illustrated in the previous section, where it was shown that the analysis of the convergence of Algorithm 1 could be reduced to the optimality of stable or strongly stable solutions. The convergence of the algorithm proposed in GARCÍA [2, 3] was also analyzed there with the help of these theoretical results. It is important to note that, as we pointed out in Section 5, the conditions of equilibrium in Theorem 3 are in general not sufficient conditions of optimality. Therefore, if the partition $\mathcal{Q}$ is fixed, hypothesis (iv) of Theorems 2 and 4 is an essential one.

References

[1] García, L.: Sobre la agregación de modelos lineales. Investigación Operacional, **15** (1975) 30–46.

[2] García, L.: Algoritmo iterativo de agregación para el problema general de programación lineal. Investigatión Operacional, **IX** (1988) 3 (to appear).

[3] García, L.: An iterative aggregation algorithm for linear programming. (To appear at the proceedings of the Seminar on Approximation and Optimization, Havana, Cuba, January 1987).

[4] García, L.: Estudio de una clase de algoritmos de agregación. (Ph. D. Dissertation, Academy of Sciences, Cuba, 1987).

[5] Luenberger, D. G.: Introduction to Linear and Nonlinear Programming. Addison-Wesley, California, 1973.

[6] Martínez, F.: Método iterativo de agregación en programación lineal. Investigación Operacional, **11** (1974) 19–25.

[7] Mendelssohn, R.: Improved bounds for aggregate Linear programs. Operations Research, **28** (1980) 1450–1453.

[8] Mendelssohn, R.: An iterative aggregation procedure for Markov decision processes. Operations Research, **30** (1982) 1, 62–72.

[9] Shapiro, J. F.: Mathematical Programming: Structures and Algorithms. New York, John Wiley, 1979.

[10] Vakhutinsky, I. Y., L. M. Dudkin: Algorithm of iterative aggregation for the solution of the problem of Linear Programming of a general nature. Inzvestia Novosibirski, Seria Obshchestvendykh Nauk, **11** (1973) 67–71.

[11] Vakhutinsky, I. Y., L. M. Dudkin, A. A. Yvkin: Iterative aggregation. A new approach to the solution of large-scale problems. Econometrica, **47** (1979) 821–841.

[12] Zipkin, P.: Bounds on the effect of aggregating variables in linear programs. Operations Research, **28** (1980) 403–418.

Received April 86, revised December 87

L. García
Inst. de Cibernética, Matemática y Física, Academia de Ciencias
Calle 0 # 8
Habana 4
Cuba

optimization **19** (1988) 6, 829–838

An Exact Penalty Approach for Solving a Class of Minimization Problems with Boolean Variables

MARLIES BORCHARDT

Technische Hochschule Ilmenau, Sektion Mathematik,
Rechentechnik und ökonomische Kybernetik

Summary: An exact penalty approach for solving minimization problems with a concave objective function, linear constraints and Boolean variables is proposed. The penalty problems have continuous variables. An estimation of the penalty parameter which guarantees the exactness can be calculated on the base of an auxiliary problem. The results are applied to problems with an arbitrary quadratic objective function, linear constraints and Boolean variables. This leads to a modified Lagrangian approach for the latter problems. In the general case, the penalty approach is compared with a direct application of results of global optimization to a modification of the initial problem.

AMS 1980 Subject Classification:

Primary: 90 C 09; Secondary: 90 C 30

Key words: Boolean programming, exact penalty method, modified Lagrangian.

1. Introduction

In the last 20 years, a lot of papers was devoted to Boolean optimization problems. HANSEN [5] reported about the development of nonlinear Boolean programming until 1978. Since this time, many authors published articles which deal with the special case where the objective function is quadratic and there are either none constraints or the constraints are linear or quadratic [8], [11] [(see [3] for further references).

Linear problems with Boolean variables are often solved efficiently by linear programming based branch-and-bound algorithms. But KALANTARI/ROSEN [7] have shown the possibility of introducing a penalty problem where a solution of this one implies a solution of the original linear Boolean problem.

HANSEN [5] has noted that any nonlinear 0–1 program is equivalent to a polynomial 0–1 knapsack problem in the same variables. The work necessary to obtain this single constraint may, however, be as large as that required to solve the initial problem directly (see also [3]). He reported that any bounded nonlinear Boolean problem with integer valued equality constraints can be solved by a penalty approach. But his approach is connected with a large augmentation of the numbers of terms and variables.

In [1], [2] we proposed a penalty approach for solving problems with mixed discrete and continuous variables. We showed that there exists a finite penalty parameter such that an optimal point of the corresponding penalty problem implies the optimal values of the discrete variables of the initial problem if the feasible set of this one is not empty. The penalty problems are such ones with mixed variables.

Now we intend to give an exact penalty approach for solving a class of Boolean problems where the drawbacks of the method of HANSEN not occur. We show that an optimal point of such a problem is also optimal to the initial problem if the penalty parameter is chosen sufficiently large but finite.

Further, we give an estimation to the penalty parameter needed to guarantee the effect mentioned above. This parameter can be calculated in advance by finding the solution of an auxiliary problem. We remember that the multiplier methods in nonconvex programming caused by HESTENES [6] and ROCKAFELLAR [10] provide (under some regularity assumptions) the solution of the original problem via surrogate problems defined by a modified Lagranian if the associated penalty parameter is sufficiently large but finite. An estimation of the value of the penalty parameter which ensures this effect was given some years later by CIRINA [4].

The concept of exact penalty functions in nonlinear continuous optimization is due to PIETRZYKOWSKI [9]. To our knowledge the penalty parameter which guarantees the exactness can't be calculated in advance.

Our paper is organized as follows. We formulate the initial problem and associate a penalty problem in Section 2. In Section 3, we show that T is an exact penalty function to (P), that means there exists a penalty parameter $\bar{r}$ such that a solution $x(r)$ of (SP(r)) implies a solution of (P) whenever $r \geqq \bar{r}$, i.e. we needn't solve the discrete problem (P) but we have to solve the continuous problem (SP(r)). The question how to solve it we discuss in Section 4. The penalty parameter can be calculated in advance by an auxiliary problem. We give a lower estimation of this parameter. As examples show, this lower estimation guarantees sometimes the exactness effect. In Section 5 we apply our results to problems with arbitrary quadratic objective functions. Hereby we give an interpretation in the context of modified Lagrangians.

In Section 6, we compare our approach with a direct application of methods of global optimization to our general initial problem. We show that our penalty technique can be applied to advantage.

2. Formulation of the Problem

We consider the following problem

(P) $\quad \mathrm{MIN}\,\{f(x) \mid x \in G\}$

where G is given by

$$G := \{x \in B_2^n \mid Ax \leqq b\}\,, \tag{2.1}$$
$$B_2^n := \{x \in \boldsymbol{R}^n \mid x_i \in \{0, 1\},\ i = 1, \ldots, n\}\,,$$

$f: M \to \boldsymbol{R}$ is a concave function and $M \supseteqq \operatorname{co} B_2^n$. $\operatorname{co} B_2^n$ denotes the convex hull of B_2^n, A is an (m, n)-matrix and $b \in \boldsymbol{R}^m$. Generally, we assume $G \neq \emptyset$.

Further we introduce the set

$$\bar{G} := \{x \in \operatorname{co} B_2^n \mid Ax \leqq b\} . \tag{2.2}$$

$\bar{G}$ is a closed bounded polyhedron. The problem

(cR) $\quad \operatorname{MIN} \{f(x) \mid x \in \bar{G}\}$

is called the continuous relaxation of (P).

It is known, that the conditions $x_i \in \{0, 1\}$, $i = 1, \ldots, n$, can be equivalently written as

$$x_i^2 - x_i = 0, \quad i = 1, \ldots, n . \tag{2.3}$$

Let be $e := (1, 1, \ldots, 1) \in \boldsymbol{R}^n$. Furthermore, we define

$$Q := \{x \in \boldsymbol{R}^n \mid \langle x, e - x \rangle = 0\}$$

and get

$$G = \bar{G} \cap Q .$$

We consider the problem

(SP(r)) $\quad \operatorname{MIN} \{T(x, r) \mid x \in \bar{G}\}, \quad r > 0,$

$$T(x, r) := f(x) + r \langle x, e - x \rangle . \tag{2.4}$$

3. Relations between (P) and (SP(r))

Let be $E(\bar{G})$ the set of all vertices of $\bar{G}$, $\tilde{E}(\bar{G})$ the set of the vertices of $\bar{G}$ which have at least one non-integer coordinate. We regard the following convex function

$$\alpha(x) := \sum_{i=1}^{n} \left(x_i - \frac{1}{2}\right)^2 . \tag{3.1}$$

Obviously,

$$\alpha(x) = \frac{1}{4} n \quad \text{if} \quad x \in G , \tag{3.2}$$

$$\alpha(x) < \frac{1}{4} n \quad \text{if} \quad x \in \tilde{E}(\bar{G}) . \tag{3.3}$$

Introducing the following "distance"-problem

(DP) $\quad \operatorname{MAX} \{\alpha(x) \mid x \in \tilde{E}(\bar{G})\}$

we look for the non-integer vertex of $\bar{G}$ which is farthest from the center $\left(\frac{1}{2}, \frac{1}{2}, \ldots, \frac{1}{2}\right) \in \boldsymbol{R}^n$ of $\operatorname{co} B_2^n$. $\bar{G}$ is a closed bounded polyhedron. Therefore, the problem (DP) possesses at least one solution x^*:

$$\alpha(x^*) \geqq \alpha(x) \quad \forall x \in \tilde{E}(\bar{G}) . \tag{3.4}$$

Besides $(\mathrm{SP}(r))$, we consider the following modified penalty problem

$$(\overline{\mathrm{SP}}(r)) \qquad \mathrm{MIN}\,\{S(x, r) \mid x \in \bar{G}\}\,,$$
$$S(x, r) := f(x) - r\alpha(x)\,. \tag{3.5}$$

It holds

$$T(x, r) = S(x, r) + \frac{1}{4}\, n \cdot r\,, \tag{3.6}$$

Theorem 1: *Let be f concave, x^* a solution of* (DP), *x^0 a solution of* (cR) *and*

$$f(x) \leqq 0 \quad \forall x \in G\,. \tag{3.7}$$

If $x(r)$ is a solution of

$$(\overline{\mathrm{SP}}(r)) \qquad \mathrm{MIN}\,\{S(x, r) \mid x \in \bar{G}\}$$

where S is defined in (3.5) *and*

$$r > \bar{r} := \frac{-f(x^0)}{\left(\frac{1}{4}\, n - \alpha(x^*)\right)}\,, \tag{3.8}$$

then $x(r)$ is a solution of (P).

Proof: (3.8), (3.7), (3.3) and $x^* \in \tilde{E}(\bar{G})$ imply $\bar{r} > 0$. α defined accordingly (3.1) is a convex function. Therefore the concavity of f and $r \geqq 0$ imply that $S(\cdot, r)$ is concave. Since the minimum of a concave function over a closed and bounded polyhedron is attained at one of its extreme points we have

$$\min_{x \in \bar{G}} S(x, r) = \min_{x \in E(\bar{G})} S(x, r) \quad \forall r > \bar{r}\,. \tag{3.9}$$

Let be $x(r)$ a solution of $(\overline{\mathrm{SP}}(r))$ with r accordingly (3.8), that means

$$x(r) \in \bar{G}, \quad S(x(r), r) \leqq S(x, r) \quad \forall x \in \bar{G}\,. \tag{3.10}$$

We assume

$$x(r) \in \tilde{E}(\bar{G})\,, \tag{3.11}$$

and choose some $\bar{x} \in G$. Then, we have $\alpha(\bar{x}) = \frac{1}{4}\, n$ and obtain

$$S(\bar{x}, r) - S(x(r), r) = f(\bar{x}) - f(x(r)) - r\left(\frac{1}{4}\, n - \alpha(x(r))\right). \tag{3.12}$$

Because of $x(r) \in \bar{G}$ it follows from (3.4) the relation

$$\alpha(x^*) \geqq \alpha(x(r))\,.$$

Hence, (3.12) provides

$$S(\bar{x}, r) - S(x(r), r) \leqq f(\bar{x}) - f(x(r)) - r\left(\frac{1}{4}\, n - \alpha(x^*)\right). \tag{3.13}$$

(3.8) implies

$$S(\bar{x}, r) - S(x(r), r) < f(\bar{x}) - f(x(r)) + f(x^0)$$

and it follows with (3.7) the relation

$$S(\bar{x}, r) - S(x(r), r) < f(x^0) - f(x(r)) . \tag{3.14}$$

x^0 is a solution of (cR) and $x(r) \in \bar{G}$, thus we have

$$f(x^0) \leqq f(x(r)) .$$

Therefore, we obtain from (3.14) the relation

$$S(\bar{x}, r) < S(x(r), r)$$

which contradicts (3.10), i.e. our assumption (3.11) was not correct. Now, we get with (3.9) the relation $x(r) \in E(\bar{G}) \setminus \tilde{E}(\bar{G})$ and consequently,

$$x(r) \in G . \tag{3.15}$$

(3.5), (3.2) and (3.15) provide

$$S(x(r), r) = f(x(r)) - \frac{1}{4} nr$$

and

$$f(x(r)) \leqq f(x) \quad \forall x \in G. \quad \blacksquare$$

The assumption (3.7) doesn't cause any losse of generality because this condition can be realized by a translation whenever f is finite on G, i.e. we can replace the condition (3.7) of Theorem 1 by

$$f \text{ is finite on } G . \tag{3.16}$$

Theorem 1 and (3.6) show that the problems $(\mathrm{SP}(r))$ and $(\overline{\mathrm{SP}}(r))$ have the same solution but the optimal values of these problems differ by $\frac{1}{4} nr$. Thus we can formulate the following main result:

Theorem 2: *Let be the assumption of Theorem 1 satisfied. If $x(r)$ is a solution of $(\mathrm{SP}(r))$ with r according to (3.8) then $x(r)$ is a solution of (P).*

Example 1: We consider the problem

$$\mathrm{MIN} \left\{ -x^2 \mid x \in \{0, 1\}, x \leqq \frac{1}{3} \right\}.$$

The solution of this problem is $\tilde{x} = 0$. The solution of (cR) is $x^0 = \frac{1}{3}$. We have $\tilde{E}(\bar{G}) = \left\{ \frac{1}{3} \right\}$, $x^* = \frac{1}{3}$, $\alpha(x^*) = \frac{1}{36}$. Further we get $\bar{r} = \frac{1}{2}$. The penalty problem

$$(\mathrm{SP}(r)) \qquad \mathrm{MIN} \left\{ -x^2 + \frac{1}{2} x \cdot (1 - x) \mid 0 \leqq x \leqq \frac{1}{3} \right\},$$

has the solution $x(r) = 0$ if $r > \frac{1}{2} (= \bar{r})$. •

4. On the Solution of the Surrogate Problems

Numerically, it is difficult to find x^* as solution of (DP). As a lower estimation of $\bar{r}$ we can use

$$r_0 := \frac{-f(x^0)}{\left(\frac{1}{4} n - \alpha(x^*)\right)} \tag{4.1}$$

if $x^0 \in \tilde{E}(\bar{G})$. In fact, x^0 belongs to $E(\bar{G})$ because x^0 is a solution of (cR) which is a problem of minimizing a concave function over a closed bounded polyhedron. If $x^0 \notin \tilde{E}(\bar{G})$ then we have already $x^0 \in G$ and are ready. x^0 is a solution of (P). Therefore, we have only to consider the case $x^0 \in \tilde{E}(\bar{G})$. (3.4) implies

$$\alpha(x^*) \geqq \alpha(x^0) \tag{4.2}$$

and we get from (4.1), (4.2) and (3.8) the relation

$$r_0 \leqq \frac{-f(x^0)}{\left(\frac{1}{4} n - \alpha(x^*)\right)} = \bar{r} .$$

Starting with r_0 we can apply the usual penalty strategy, i.e. we solve the problems (SP(r)) for a sequence of penalty parameters $\{r_k\} \subseteq (r_0, \tilde{r})$ with $\tilde{r} \geqq \bar{r}$. If $x(r) \in G$ then we are ready. Because of Theorem 2 this relation is satisfied if r is sufficiently large but finite. As examples show, this relation can sometimes be satisfied if $r = r_0$. This is especially the case if $\tilde{E}(\bar{G}) = \{x^0\}$. The following example illustrates that also in the case $r_0 < \bar{r}$ a solution $\tilde{x}$ of (P) can be implied by $x(r_0)$.

Example 2: Now we consider the problem

$$\text{MIN } \{-x_1^2 \mid x_1 + 4x_2 \leqq 3,\ x_1 \in \{0, 1\},\ x_2 \in \{0, 1\}\} .$$

Its solution is $\tilde{x} = (1, 0)$. We get $x^0 = \left(1, \frac{1}{2}\right)$ $f(x^0) = -1$,

$$x^* = \left(0, \frac{3}{4}\right), \quad \alpha(x^*) = \frac{5}{16}, \quad \alpha(x^0) = \frac{1}{4}$$

and $\bar{r} = \frac{16}{3}$, $r_0 = 4$. We obtain $x(r_0) = (1, 0)$ and therefore $x(r_0) = \tilde{x}$.

The penalty problems (SP(r)) and the problem (cR) are problems of minimizing a concave function over a closed bounded polyhedron. Therefore we can use schemes of Hoang Tuy [12], [13], [14] to solve these problems. These methods provide a solution of (SP(r)) after a finite number of steps. So, a solution of the initial problem is also reached after a finite number of steps.

5. Application of the Results to Problems with Arbitrary Quadratic Objective Functions

We consider the problem

(PQ) $\quad \text{MIN } \{\langle x, Cx\rangle + \langle c, x\rangle \mid x \in G\}$

where C is a symmetric (n, n)-matrix, $c \in \boldsymbol{R}^n$ and G is given by (2.1). (2.3) implies that (PQ) is equivalent to

$$(\mathrm{PQ}(y)) \qquad \mathrm{MIN}\,\{\langle x, (C-y)\,x\rangle + \langle c+y, x\rangle \mid x\in G\}$$

where $y := (y_1, \dots, y_n)\in \boldsymbol{R}^n$ and Y is an (n, n)-diagonal matrix with elements y_i, $i=1, \dots, n$ at the diagonal. If the elements y_i, $i=1, \dots, n$, are chosen sufficiently large the objective function of $(\mathrm{PQ}(y))$ is concave. A suitable y can be found by a modified CHOLESKY-method. Now we can apply Theorem 2 to $(\mathrm{PQ}(y))$ where we replace (3.7) by (3.16).

Theorem 3: *Let be $x(y, r)$ a solution of*

$$(\mathrm{SPQ}(y, r))\ \mathrm{MIN}\,\{\langle x, (C-Y)\,x\rangle + \langle c+y, x\rangle + r\,\langle x, e-x\rangle \mid x\in \bar{G}\}$$

where $\bar{G}$ is given by (2.2). Then there exist $\bar{y}\in\boldsymbol{R}^m_+$ and $\bar{\bar{r}}\in\boldsymbol{R}_+\{0\}$ (both finite) such that $x(y, r)$ is a solution of (PQ) whenever $y\geqq\bar{y}$ and $r\geqq\bar{\bar{r}}$.

An estimation of $\bar{\bar{r}}$ can be calculated in the same way as above (see (13.8) and (4.1)), but $\bar{\bar{r}}$ depends on $\bar{y}$. The objective function $L:\boldsymbol{R}^n\times\boldsymbol{R}^n\times\boldsymbol{R}_+\to\boldsymbol{R}$,

$$L(x, y, r) := \langle x, (C-Y)\,x\rangle + \langle c+y, x\rangle + r\,\langle x, e-x\rangle\,,$$

can be interpreted as a modified Lagrangian of (PQ). Obviously,

$$\begin{aligned} L(x, y, r) &= \langle x, Cx\rangle + \langle c, x\rangle + \sum_{i=1}^{n} y_i\,(x_i^2 - x_i) + r\,\langle x, e-x\rangle \\ &= L_0(x, y) + r\,\langle x, e-x\rangle \end{aligned}$$

where

$$L_0(x, y) := \langle x, Cx\rangle + \langle c, x\rangle + \sum_{i=1}^{n} y_i\,(x_i^2 - x_i)$$

is the classical Lagrangian of (PQ) where only the equality constraints $x_i^2 - x_i = 0$ are dualized (cf. (2.1), (2.3)). The term $\langle x, e-x\rangle$ is a penalty term to (PQ) where also only the equality constraints of (PQ) are taken under consideration. In this sense, the problem $(\mathrm{SPQ}(y, r))$ can be taken as a (modified) Lagrangian relaxation of the original problem (PQ). The result of Theorem 3 means that a solution of the Lagrangian relaxation $(\mathrm{SPQ}(y, r))$ implies a solution of the original problem (PQ) if the parameters y and r are chosen sufficiently large. In this form, it has analogies in nonlinear continuous based on other modifications of the classical Lagrangian.

6. Comparison of a Direct Approach for Solving (P) with the Penalty Approach

(P) can equivalently be written as

$$(\mathrm{P}) \qquad \mathrm{MIN}\,\{f(x) \mid Ax\leqq b,\ x\in \mathrm{co}\,B_2^n,\ x_i^2 - x_i \geqq 0,\ i=1, \dots, n\}\,.$$

The constraints $x_i^2 - x_i \geqq 0$, $i=1, \dots, n$, are revers—convex. (P′) can be transformed in a so called canonical d.c.-problem of the form

$$(\tilde{\mathrm{P}}) \qquad \mathrm{MIN}\,\{\langle c, z\rangle \mid z\in\Omega,\ H(z)\geqq 0\}\,,$$

where $c \in \boldsymbol{R}^{n+2}$, $\Omega \subseteqq \boldsymbol{R}^{n+2}$ is a closed convex set and $H : \boldsymbol{R}^{n+2} \to \boldsymbol{R}$ is a convex function.

(P′) can be written as

(P″) $\quad \text{MIN } \{t \mid Ax \leqq b,\ x \in \text{co } B_2^n,\ x_i^2 - x_i \geqq 0,\ i=1, \ldots, n,\ -f(x) \geqq -t,\ t \in \boldsymbol{R}\}$

with $(n+1)$ revers-convex constraints. To solve this problem with methods of TUY [13], [14] we have to transform it in a problem with only one revers-convex constraint. This can be done by summarizing these $(n+1)$ constraints accordingly

$$G(x, t) := \min_{i \in \{1, \ldots, n\}} G_i(x, t) \geqq 0 \,, \tag{6.1}$$

$$G_i(x, t) := x_i^2 - x_i, \quad i = 1, \ldots, n \,, \tag{6.2}$$

$$G_{n+1}(x, t) := t - f(x) \,. \tag{6.3}$$

G is a d.c. function. It can be written as the difference of two convex functions

$$G(x, t) = \sum_{i=1}^{n+1} G_i(x, t) - \max_{j \in \{1, \ldots, n+1\}} \sum_{\substack{i=1 \\ i \neq j}}^{n+1} G_i(x, t) \,. \tag{6.4}$$

Following [14] we set

$$H_1(x, t) := \sum_{i=1}^{n+1} G_i(x, t) \,, \tag{6.5}$$

$$H_2(x, t) := \max_{j \in \{1, \ldots, n+1\}} \sum_{\substack{i=1 \\ i \neq j}}^{n+1} G_i(x, t) \,. \tag{6.6}$$

From (6.1), (6.5) we get

$$G(x, t) = H_1(x, t) - H_2(x, t) \geqq 0 \,. \tag{6.7}$$

But (6.7) is equivalent to

$$H_1(x, t) - \alpha \geqq 0 \tag{6.8}$$

$$\alpha - H_2(x, t) \geqq 0 \tag{6.9}$$

where H_1, H_2 are convex functions.

(6.2), (6.3), (6.5) and (6.6) imply in our special case

$$H_1(x, t) = \sum_{i=1}^{n} (x_i^2 - x_i) + t - f(x) \,,$$

$$H_2(x, t) = \max \left\{ \max_{j \in \{1, \ldots, n\}} \left\{ \sum_{\substack{i=1 \\ i \neq j}}^{n} (x_i^2 - x_i) + t - f(x) \right\}, \sum_{i=1}^{n} (x_i^2 - x_i) \right\} .$$

Now we can transform (P″) to

(P‴) $\quad \text{MIN } \left\{ t \mid Ax \leqq b,\ x \in \text{co } B_2^n,\ \max \left\{ \max_{j \in \{1, \ldots, n\}} \left\{ \sum_{\substack{i=1 \\ i \neq j}}^{n} (x_i^2 - x_i) + t - f(x) \right\}, \right.\right.$

$\left.\left. \sum_{i=1}^{n} (x_i^2 - x_i) \right\} - \alpha \leqq 0,\ \sum_{i=1}^{n} (x_i^2 - x_i) + t - f(x) - \alpha \geqq 0,\ \alpha \in \boldsymbol{R},\ t \in \boldsymbol{R} \right\}.$

We set

$$\Omega := \Big\{ z := (x, t, \alpha) \in \boldsymbol{R}^{n+2} \mid Ax \leqq b,\ x \in \operatorname{co} B_2^n,\ \alpha \in \boldsymbol{R},\ t \in \boldsymbol{R},$$

$$\max \Big\{ \max_{j \in \{1,\ldots,n\}} \Big\{ \sum_{\substack{i=1 \\ i \neq j}}^{n} (x_i^2 + x_i) + t - f(x) \Big\},\ \sum_{i=1}^{n} (x_i^2 - x_i) \Big\} - \alpha \leqq 0 \Big\}, \tag{6.10}$$

$$H(z) := \sum_{i=1}^{n} (x_i^2 - x_i) + t - f(x) - \alpha = \langle x, x - e \rangle + t - f(x) - \alpha, \tag{6.11}$$

$$c := (0, \ldots, 0, 1, 0) \in \boldsymbol{R}^{n+2}$$

and get (P‴) in the form $(\tilde{\mathrm{P}})$.

The canonical d.c. form of $(\mathrm{SP}(r))$ is

$(\widetilde{\mathrm{SP}(r)})$ $\quad \mathrm{MIN}\, \{ t \mid x \in \operatorname{co} B_2^n,\ Ax \leqq b,\ t - f(x) + r \langle x, e - x \rangle \geqq 0,\ t \in \boldsymbol{R} \}$

where r is chosen fixed. In this case, we have

$$\Omega := \{ (x, t) \in \boldsymbol{R}^{n+1} \mid x \in \operatorname{co} B_2^n,\ Ax \leqq b,\ t \in \boldsymbol{R} \} \tag{6.12}$$

and

$$H(x, t) := t - f(x) + r \langle x, x - e \rangle .$$

Ω accordingly (6.12) has a much easier structure than Ω accordingly (6.10). So, the problem $(\mathrm{SP}(r))$ can be easier solved than $(\tilde{\mathrm{P}})$ with Ω and H accordingly (6.10) and (6.11) respectively. We remark that $(\tilde{\mathrm{P}})$ with Ω and H accordingly (6.10), (6.11) can be written as

$(\tilde{\tilde{\mathrm{P}}})$ $\quad \mathrm{MIN}\, \Big\{ f(x) + \langle x, e - x \rangle + \alpha \mid x \in \operatorname{co} B_2^n,\ \alpha \in \boldsymbol{R},\ t \in \boldsymbol{R},\ Ax \leqq b,$

$$\max \Big\{ \max_{j \in \{1,\ldots,n\}} \Big\{ \sum_{\substack{i=1 \\ i \neq j}}^{n} (x_i^2 - x_i) + t - f(x) \Big\},\ \sum_{i=1}^{n} (x_i^2 - x_i) \Big\} - \alpha \leqq 0 \Big\} .$$

That means, we have to minimize a concave function over a convex set. But if we consider $(\mathrm{SP}(r))$, we have to minimize the concave function $f(x) + r \langle x, e - x \rangle$ over the closed bounded polyhedron

$$\{ x \in \operatorname{co} B_2^n \mid Ax \leqq b \} .$$

But this latter problem can be easier solved for example by a method given by TUY [12].

References

[1] BORCHARDT, M.: Straf- und Multiplikatormethoden für Optimierungsprobleme mit diskreten und stetigen Variablen (I). Wiss. Z. TH Ilmenau, **31** (1985) 4, 51–58.

[2] BORCHARDT, M.: Straf- und Multiplikatormethoden für Optimierungsprobleme mit diskreten und stetigen Variablen (II). Wiss. Z. TH Ilmenau, **32** (1986) 1, 63–72.

[3] BORCHARDT, M.: Zur Lösung von nichtlinearen Booleschen Optimierungsproblemen. Wiss. Z. TH Ilmenau, **33** (1987) 4, 101–111.

[4] CIRINA: A remark on multiplier methods for nonlinear programming. In: Lecture Notes in Computer Sciences, Vol. 41 (1977). Series I.F.I.P. TC 7, Optimization Conference, 283–292.

[5] HANSEN, P.: Methods of nonlinear 0-1 programming. In: HAMMER, P. L.; E. L. JOHNSON; B. H. KORTE (eds.): Discrete optimization II. Annals of discrete mathematics. North Holland Publ.-Company-Amsterdam, New York, Oxford 1979, 53–70.

[6] HESTENES, M. R.: Multiplier and gradient methods. J. Optimization Theory Appl., **4** (1969) 5, 303–320.

[7] KALANTARI, B.; J. B. ROSEN: Penalty for zero-one integer equivalent problems. Mathematical programming, **24** (1982), 229–232.

[8] KÖRNER, F.; C. RICHTER (1982): Zur effektiven Lösung von Booleschen quadratischen Optimierungsproblemen. Num. Math., **40** (1982) 99–109.

[9] PIETRZYKOWSKI, T.: An exact potential method for constrained maxima. SIAM J. Num. Anal., **6** (1969) 2, 299–304.

[10] ROCKAFELLAR, R. T.: Augmented LAGRANGE-multiplier functions and duality in nonconvex programming. SIAM J. Control, **12** (1974) 268–285.

[11] SHOR, N. Z.; A. S. DAVYDOV (1985): O metode polučenija ocenok v kvadratičnych ėkstremal'nych zadačach s bulevymi peremennymi Kibernetika (1985) 2, 48–50.

[12] TUY, H.: Vognutoje programmirovanija s linejnymi ograničenijami. Doklady Akad. Nauk SSSR, **159** (1964), 32–35.

[13] TUY, H.: Global maximization of a convex function over a closed convex, not necessarily bounded set. Centre de recherche de mathematiques de la decision n⁰ 8223.

[14] TUY, H.: A general deterministic approach to global optimization via d.c. programming. In: HIRIART-URRUTY, J.-B. (ed.): FERMAT-Days 1985: Mathematics for optimization. Elseviere Science Publ. 1986, 273–304.

Received March 1987, revised March 1988

M. Borchardt
Technische Hochschule Ilmenau
Sektion Mathematik, Rechentechnik
und ökonomische Kybernetik
Am Ehrenberg
DDR - 6300 Ilmenau

optimization 19 (1988) 6, 839–860

A New Algorithm for the Solution of the Linear Minimax Approximation Problem

N. Papamarkos and G. Vachtsevanos

Democritus University of Thrace, School of Engineering, Xanthi
School of Georgia Institute of Technology, School of Electrical Engineering, Atlanta

Summary: This paper introduces an efficient approach to the solution of the linear minimax approximation problem. The classical nonlinear minimax problem is cast into a linear formulation. The proposed optimization procedure consists of specifying first a feasible point belonging to the feasible boundary surface. Next, feasible directions of decreasing values of the objective function are determined. The algorithm proceeds iteratively and terminates when the absolute minimum value of the objective function is reached. The initial point may be selected arbitrarily or it may be optimally determined through a linear method to speed up algorithmic convergence. The algorithm was applied to a number of approximation problems and results were compared to those derived using the revised simplex method. The new algorithm is shown to speed up the problem solution by at least on order of magnitude.

AMS 1980 Subject Classification:

Primary: 49 A 40; Secondary: 90 C 05, 90 C 06.

Key words: Minimax approximation, linear programming, combinatorial optimization, parameter estimation.

1. Introduction

It is well known that many system design and parameter estimation problems may be formulated as optimization problems in the minimax sense [1–3]. Such problems are, in turn, solved by exploiting basic Linear Programming (LP) methods [4–5] and more specifically the Revised Simplex Algorithm (RSA) [6]. This procedure is satisfactory for problems of small dimensionality (as it refers to the number of parameters and the aggregate of constraints), but its utility for problems of large dimensionality is questionable since it exhibits such undesirable characteristics as probabilistic data storage requirements, poor speed of convergence and high roundoff errors resulting from a continuous inversion of the base as demanded by RSA.

The same problem may be considered as a special case of nonlinear optimization methods; the penalty function [7, 8] and the nonlinear minimax optimization problem [9, 10] belong to this category. These algorithms are difficult, though, to implement, and involve a higher computational cost than linear programming methods. Finally, they do not always guarantee convergence to the global minimum.

The general formulation of the minimax problem consists of determining the variables x_j, $j=1, 2, \ldots, N$ through the solution of the problem

minimize ξ

under the constraints

$$\left| \sum_{j=1}^{N} d_{ij} x_j - p_i \right| \leqq \xi \quad \textit{for} \quad i=1, 2, \ldots, M \tag{1}$$

$\xi \geqq 0$, x_j, $j=1, 2, \ldots, N$. The x_j variables are unresricted in sign and d_{ij}, p_i are known coefficients.

Without loss of generality, it is considered that $p_i \geqq 0$ $\forall i=1, 2, \ldots, M$.

The nonlinear problem (1) is converted to the following linear form by dropping the absolute values:

minimize ξ

under the constraints

$$-\xi + \sum_{j=1}^{N} d_{ij} x_j - p_i \leqq 0$$

$$-\xi - \sum_{j=1}^{N} d_{ij} x_j + p_i \leqq 0 \quad \textit{for} \quad i=1, 2, \ldots, M\,. \tag{2}$$

This new formulation (2) contains $N+1$ variables and $2M$ constraints. For its solution, using the revised Simplex algorithm, the matrix of the d_{ij} coefficients with dimensions $(2M)\times(N+1)$, as well as the array containing the base which is of dimension $(N+2)\times(N+2)$ for the primal and $(2M+2)\times(2M+2)$ for the dual problem, must be placed in storage, among other requirements for the implementation of the algorithm. It is noted that, at each iteration of the RSA the base contains the inverse matrix of the basis vectors; it is this matrix that is recomputed at each iteration step. When the dimensionality of the base is large the storage requirements present a formidable task while roundoff errors are accentuated due to the excessive number of mathematical operations. More specifically, because the base is derived, at each iteration, from its previous value, the roundoff errors tend to accumulate.

This work is introducing a new algorithm for the solution of the minimax approximation problem. While avoiding the shortcomings of the RSA approach mentioned above, it capitalizes upon the characteristic properties of the constraints and the form of the feasible region they generate. The proposed algorithm requires an initial point for its initiation, and as such any feasible point may suffice. A more judicious choice though can be made using a fast linear procedure which gives an optimum initial point close to the problem's global minimum.

Moreover, it is pointed out that the algorithm places no constraints on the sign of the problem variables in contrast to the RSA method which, as it is known, requires nonnegative values of the variables.

2. Formulation and Definitions

The problem (2) may also be transformed to the linear form of (3) using (4) and (5), i.e., forms (2) and (3) are equivalent.

minimize ξ

with constraints

$$-\xi+\sum_{j=1}^{N} a_{ij}x_j - c_i \leqq 0 \quad \textit{for} \quad i=1, 2, \ldots, 2M \tag{3}$$

where the coefficients a_{ij} *are defined by the relations*:

$$a_{ij}=\begin{cases} d_{ij} & \textit{for} \quad i=1, 2, \ldots, M \\ -d_{ij} & \textit{for} \quad i=M+1, M+2, \ldots, 2M \end{cases} \tag{4}$$

$$c_i=\begin{cases} p_i & \textit{for} \quad i=1, 2, \ldots, M \\ -p_i & \textit{for} \quad i=M+1, M+2, \ldots, 2M . \end{cases} \tag{5}$$

We define next as *intersections* ξ_k the quantities

$$\xi_k \triangleq \sum_{j=1}^{N} a_{kj}x_j - c_k, \quad \text{for} \quad k=1, 2, \ldots, 2M . \tag{6}$$

From the above definitions, it follows that $c_i \geqq 0$ for $i \leqq M$ and $c_i \leqq 0$ for $M < i \leqq 2M$ whereas the ξ_k's obey the relation

$$\xi_k = -\xi_{k+M} . \tag{7}$$

If $\boldsymbol{X}=(x_1, x_2, \ldots, x_N)^{\mathsf{T}}$ is the vector of the x_j's, a geometrical interpretation of the problem, in $N+1$ space for $M=5$, is depicted in Figure 1.

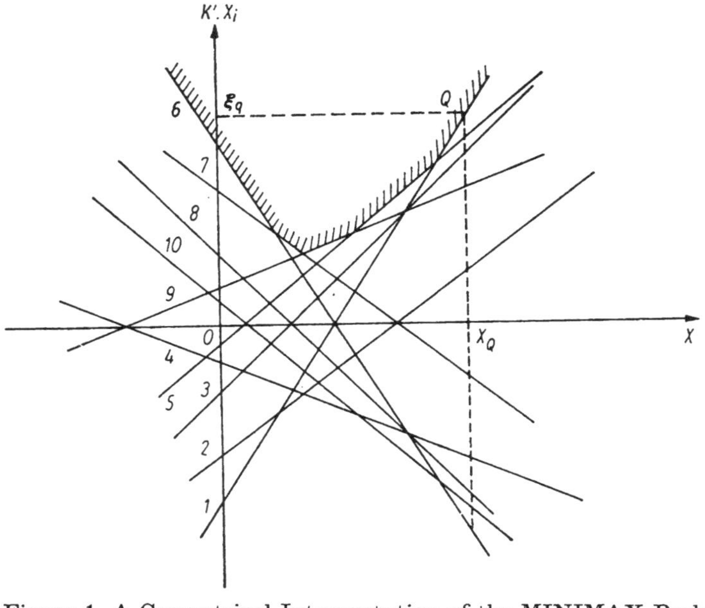

Figure 1. A Geometrical Interpretation of the MINIMAX Problem

From this figure, we conclude that:

(1) The hyperplanes i and $i+M$, defined by the constraints (6), intersect on the hyperplane of the variables $\boldsymbol{X}$.

(2) The intersections of the hyperplanes with the ξ-axis correspond to $-c_k$ for $k \leqq M$ and c_k for $M < k \leqq 2M$.

(3) The problem has a positive global minimum.

(4) For every vector $\boldsymbol{X}_k = [x_1^k, \ldots, x_N^k]^{\mathrm{T}}$, there is only one solution point on the boundary surface of the feasible region, which corresponds to ξ^k and which satisfies the condition:

$$\xi^k = \max_{i=1,2,\ldots,2M} \left(\sum_{j=1}^{N} a_{ij} x_j^k - c_i \right). \tag{8}$$

Lemma 1: *The minimax approximation problem has a convex Feasible Boundary Surface* (FBS), *which consists of all points that satisfy* (8).

The detailed proof is omitted. It can be generally stated that since the minimax approximation problem may be transformed to an LP form, then the feasible region is convex.

Lemma 2: *A line in* $N+1$ *space, defined by two points,* Q_1 *and* Q_2, *at least one* (Q_1) *of which lies on the* FBS *and on the q hyperplane, will either intersect the* FBS *at a second point on the hyperplane* $l \neq q$ *or will be identical with the initial feasible point.*

Proof: Let $(\xi^1, \boldsymbol{X}_1)$ the first feasible point on the q hyperplane, and $(\xi^2, \boldsymbol{X}_2)$ the second point not belonging to q. The line specified by these two points in $N+1$ space may be described by the equations:

$$\boldsymbol{X} = \boldsymbol{X}_1 + V\,(\boldsymbol{X}_2 - \boldsymbol{X}_1) \tag{9}$$

and

$$\xi = \xi^1 + V\,(\xi^2 - \xi^1) \tag{10}$$

where V prescribes the slope of the line. For a proof of Lemma 2, it is sufficient to show that a point (ξ^l, x_l) exists on the FBS that satisfies relations (9) and (10). For such a point, it must be true according to (8) that

$$\begin{aligned}
\xi^l &= \max_{i=1,2,\ldots,2M} \left(\sum_{j=1}^{N} a_{ij}\,(x_j^1 + V\,(x_j^2 - x_j^1)) - c_i \right) \\
&= \max_{i=1,2,\ldots,2M} \left((1-V) \sum_{j=1}^{N} (a_{ij} x_j^1 - c_i) + V \left(\sum_{i=1}^{N} a_{ij} x_j^2 - c_i \right) \right) \\
&= (1-V)\,\xi^1 + V\xi^2 = \xi^1 + V\,(\xi^2 - \xi^1)\,.
\end{aligned}$$

It is obvious, now, that the point (ξ^l, x_l) does not belong to hyperplane q_1, since in the event that it did, the second point must belong to it, too. In this case, the line would lie on the q_1 hyperplane and, as a consequence, would have had an infinite number of common points with it. Moreover, since the FBS is a continuous contour surface consisting of intersecting hyperplanes, the feasible point (ξ^l, x_l) must lie on the intersection of two neighboring hyperplanes q_1 and l. Now, because of FBS's form, it is always possible to determine point Q_2 such that $\xi_l \leqq \xi_{Q_1}$, and, therefore, the straight line segment $\overline{Q_1Q_2}$ may be characterized as specifying the direction of decreasing ξ.

In the extreme case that $\xi^1 = \xi^2$ or $V=0$, then the intersection point will coincide with the first feasible point $(\xi^1, \boldsymbol{X}_1)$. ∎

3. Description of the Algorithm

Let an initial feasible point Q exist on the hyperplane q corresponding to $\boldsymbol{X}_Q$ and $\xi^Q = \xi_q$ according to relations (6) and (8). At this feasible point, the intersections of all the hyperplanes are computed using relations (6) and (7). It is obviously true that $\xi_k \leqq \xi_q$ for $k=1, 2, ..., 2M$. If the intersection ξ_k of the kth hyperplane is equal to ξ_q, then Q will be a common point of the hyperplanes q and k. Under this interpretation, a point Q is a common point of K hyperplanes when their intersections on Q are equal.

The optimization procedure followed by the algorithm consists of specifying, for every feasible point Q that is found and belongs to FBS, a feasible line of decreasing ξ. The line is specified by two points, one of which is Q, in accordance with the procedure outlined in Lemma 2. Thus, the line of decreasing ξ will intersect a hyperplane different than q and the point of intersection will be a feasible point with a corresponding ξ value smaller than ξ_q.

Various cases are distinguished depending upon the number of hyperplanes that share Q as a common point. A different procedure is employed for each case, as it is described in the sequel.

Case 1: *Q belongs to only one hyperplane.*
In this case (Figure 2), the line of decreasing ξ is specified by Q and the point T where T is the intersection of the hyperplane q with the ξ-axis and corresponds to $\xi_T = -c_q$. The line $\overline{TQ}$ is described by the relations

$$\xi = \xi_T + V(\xi_q - \xi_T) = -c_q + V(\xi_q + c_q) \tag{11}$$

and

$$x_j = V x_j^Q, \quad \text{for} \quad j = 1, 2, ..., N\,. \tag{12}$$

The intersection of $\overline{TQ}$ with a constraint l will give a point L for which, according to (6) and (11), the following holds true

$$\sum_{j=1}^{N} a_{lj} x_j^L - c_l = -c_q + V(\xi_q + c_q) \tag{13}$$

or, via (12):

$$V \sum_{j=1}^{N} a_{lj} x_j^Q = c_l - c_q + V(\xi_q + c_q)\,. \tag{14}$$

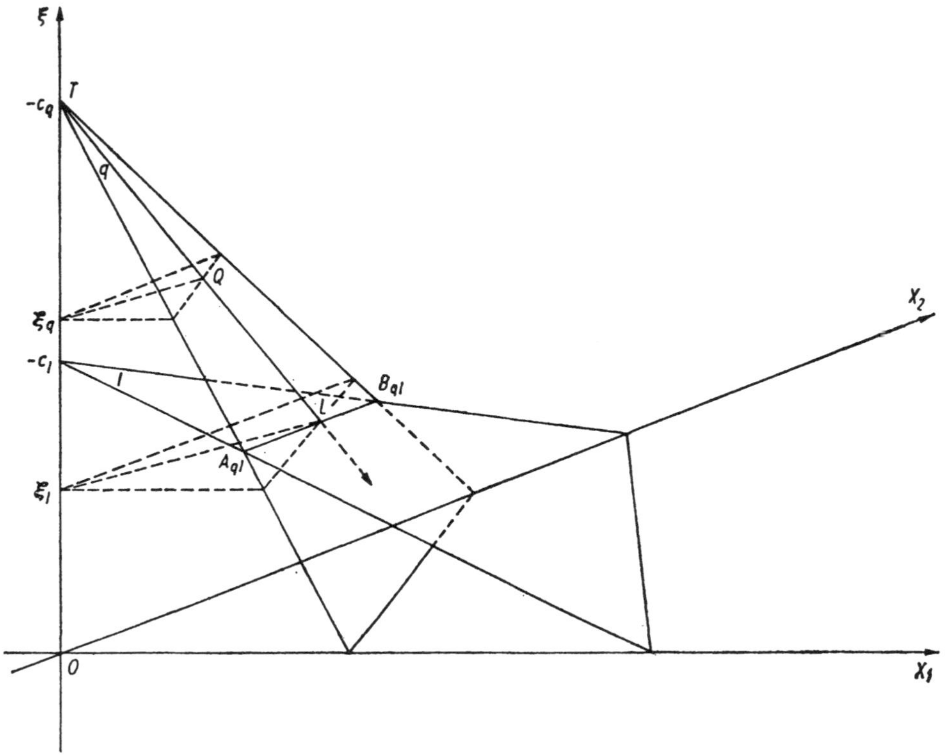

Figure 2. Feasible Direction of Decreasing ξ when Q Belongs to Only One Hyperplane

According though to (6), at the point Q it is true that

$$\sum_{j=1}^{N} a_{lj} x_j^Q = c_l + \xi_l \tag{15}$$

and relation (14) gives

$$V\ (\xi_l + c_l) = c_l - c_q + V\ (\xi_q + c_q) \tag{16}$$

and, therefore

$$V = \frac{c_l - c_q}{\xi_l + c_l - \xi_q - c_q}. \tag{17}$$

Now, knowing V, the point L may be determined through the relations (11) and (12). According to (4), the following expression describes the intersections at L:

$$\xi_k^L = \sum_{j=1}^{N} a_{kj} x_j^L - c_k$$

or, by using (12) and (6):

$$\xi_k^L = V\ (\xi_k + c_k) - c_k \quad \text{for} \quad k = 1, 2, \ldots, l-1, l+2, \ldots, 2M\,. \tag{18}$$

The above relation is important in that it permits a fast computation of the intersections at L by avoiding the calculation of the sums $\sum_j^N a_{kj} x_j^L$. Obviously and in agreement with relation (8), the point $L \neq Q$ is chosen for which $\xi_l \geqq \xi_k^l \; \forall k = 1, 2, \ldots, 2M$. It is in argument with Lemma 2 that such a feasible intersection point L exists resulting from an examination of the intersection between the feasible line in the direction of decreasing ξ with all the hyperplanes such that the above relation on ξ_l and ξ_k^l holds.

Case 2: *Q is a common point of two hyperplanes.*

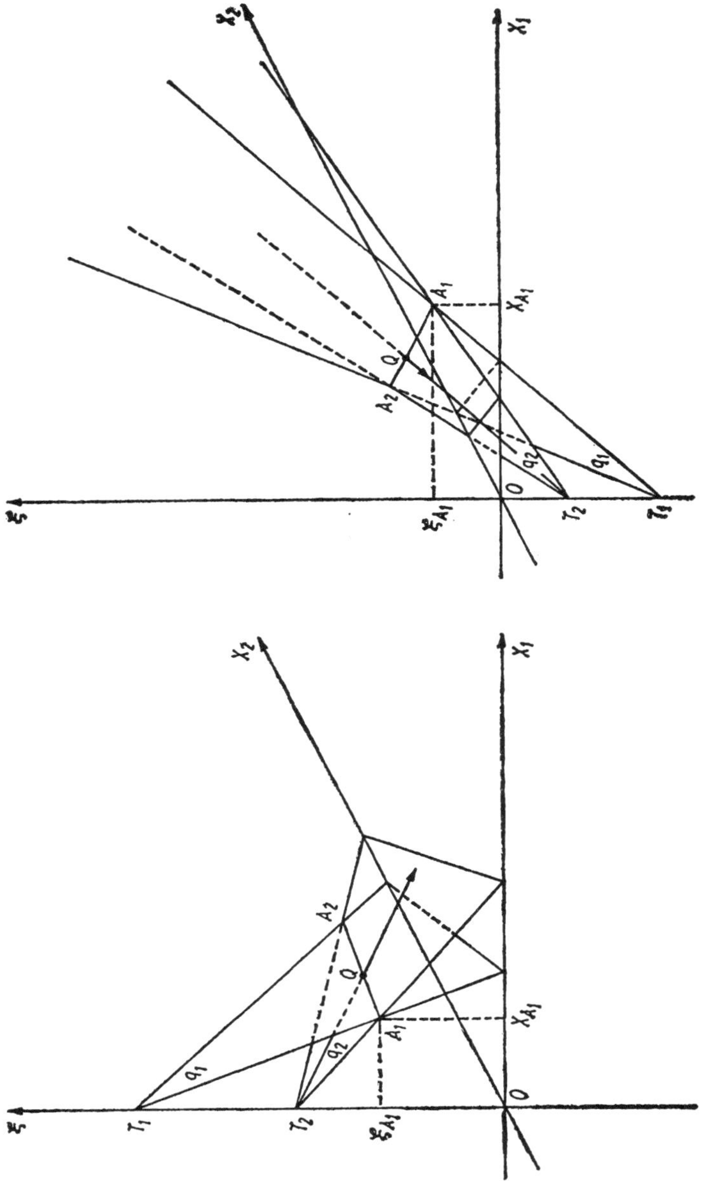

Figure 3. Two Possible Configurations when Q is a Common Point of Two Hyperplanes and $c_{q_1} c_{q_2} \geqq 0$

Let Q be a common point of two hyperplanes q_1 and q_2 and, moreover

$$|c_{q_1}| \leqq |c_{q_2}| \,. \tag{19}$$

We distinguish the following *two subcases*:

(1) $c_{q_1}c_{q_2} \geqq 0$ and the constraint $0 \leqq c_{q_1} \leqq \xi_{q_1}$ is not satisfied. Figure 3 depicts two possible configurations. Q is considered in both to lie only on hyperplane q_1 and, therefore, the procedure of Case 1, as applied to a single point, is applicable here. Obviously, hyperplane q_2 is not considered in the computation of the intersections of the feasible line in the direction of decreasing ξ, since it gives as a feasible intersection point the initial Q.

(2) $c_{q_1}c_{q_2} < 0$ or the constraint $0 \leqq c_{q_1} \leqq \xi_{q_1}$ is satisfied. In this situation, the feasible line of decreasing ξ is specified by Q and a second common point between hyperplanes q_1 and q_2. This means that the line will be part of the intersection between hyperplanes q_1 and q_{12}. The second common point, A_j, between q_1 and q_2 follows from the relations (Figure 4):

$$\begin{aligned} a_{q_1j}x_{Aj} - c_{q_1} &= \xi_{Aj} \\ a_{q_2j}x_{Aj} - c_{q_2} &= \xi_{Aj} \end{aligned} \tag{20}$$

from which it follows that

$$x_{Aj} = \frac{c_{q_1} - c_{q_2}}{a_{q_1j} - a_{q_2j}} \tag{21}$$

and

$$\xi_{Aj} = \frac{a_{q_2j}c_{q_1} - a_{q_1j}c_{q_2}}{a_{q_1j} - a_{q_2j}} \,. \tag{22}$$

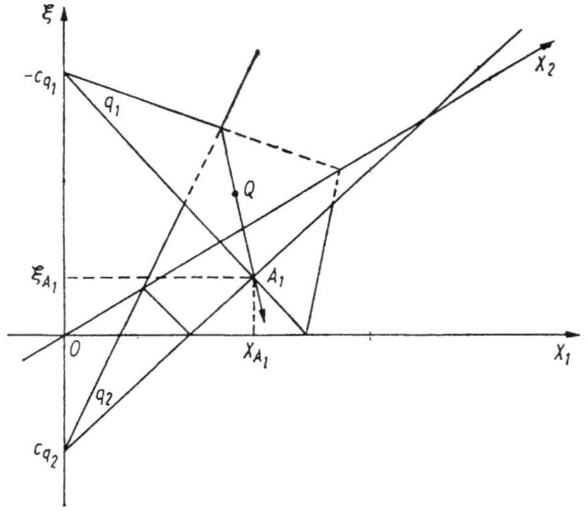

Figure 4. Q is a Common Point of Two Hyperplanes and $c_{q_1}c_{q_2} < 0$

The directed line $\overline{A_jQ}$, defined by two points according to Lemma 2, constitutes a feasible direction of decreasing ξ. It will, therefore, intersect a third hyperplane at a feasible point which will be common to three hyperplanes.

It is noted that in $(N+1)$-space, the index j in A_j corresponds to the variable x_{Aj}, which implies that there will be as many line segments $\overline{A_jQ}$ as the number of x_j variables. According to Lemma 2 though, it is sufficient for all practical purposes to specify just one such line segment $\overline{A_jQ}$.

The line segment $\overline{A_jQ}$ is described by the equations

$$\xi = \xi_{Aj} + V\ (\xi_q - \xi_{Aj}) \tag{23}$$

$$x_i = Vx_i^Q\ \forall_i \quad \text{with} \quad i \neq j \tag{24}$$

$$x_j = x_{Aj} + V\ (x_j^Q - x_{Aj})\ . \tag{25}$$

The intersection of $\overline{A_jQ}$ with a hyperplane l gives a point L. According to relations (6) and (23) it is true that

$$\xi_{Aj} + V\ (\xi_q - \xi_{Aj}) = \sum_{i=1}^{N} a_{li} x_i^L - c_l\ . \tag{26}$$

Substituting for x_i^L from relations (24) and (25):

$$\xi_{Aj} + V\ (\xi_q - \xi_{Aj}) = V \sum_{i=1}^{N} a_{li} x_i^Q + a_{lj} x_{Aj}\ (1-V) - c_l \tag{27}$$

and using expression (15), the value of V is given by

$$V = \frac{\xi_{A_j} + c_l - a_{lj} x_{A_j}}{\xi_{A_j} + c_l - a_{lj} + \xi_l - \xi_q}\ . \tag{28}$$

With V known, L may be computed through equations (23), (24) and (25). Subsequently, this point is tested using (8) to verify if it belongs to the FBS.

The intersections at L are determined by substituting the values of the variables from (24) and (25) into (6) giving

$$\xi_k^l = V \sum_{i=1}^{N} a_{ki} x_i^Q - c_k + (1-V)\ a_{kj}\ x_{Aj} \tag{29}$$

and calling again upon (6), we arrive at

$$\xi_k^l = V\ (\xi_k + c_k - a_{kj}\ x_{Aj}) + a_{kj} x_{Aj} - c_k \tag{30}$$

for $k = 1, 2, \ldots, 2M$ excluding the values $k = q_1$ and $k = q_2$. According to Lemma 2, a feasible point L always exists and it satisfies relation (8).

It is noted that the procedure of subcase (b) may be applied to subcase (a) (the inverse is not true) but it is not advisable to do so due to the increased computational cost.

Case 3: ***Q*** *is a common point of* ***T*** *hyperplanes with* $3 \leq T \leq N$.

Here, because Q is a common point of T hyperplanes in $(N+1)$-space, there will be an infinite number of common intersection points for the T hyperplanes. A feasible direction of decreasing ξ may be specified, according to Lemma 2, by a line segment through Q and a second point Q' which lies on the intersection of the T hyperplanes. That is, the line $\overline{QQ'}$ will intersect a hyperplane l at a feasible point $L \neq Q$ such that $\xi_l < \xi_q$ and L will be a common point between the T hyperplanes and l.

One method for determining the second point Q' is based on the solution of a system of linear equations. Specifically, since the corresponding value of ξ is of no consequence to Q', the following linear system of T equations is arrived at by varying both ξ_q and x_j^Q for $j = T+1, \ldots, V$:

$$\sum_{j=1}^{T} a_{ij} x_j^{Q'} = c_i + r_1 \xi_q - r_2 \sum_{j=T+1}^{N} a_{ij} x_j^Q \tag{31}$$

where the index i takes T values corresponding to T hyperplanes; and r_1, r_2 are variational coefficients. Usually $r_1 = 0$ and $r_2 > 1$.

The solution of the linear system (31) gives the $x_j^{Q'}$ values of Q' and, therefore, the line segment $\overline{QQ'}$ described by the equations:

$$\xi = \xi_q + V\,(\xi_{q'} - \xi_q) \tag{32}$$

and

$$x_j = x_i^Q + V\,(x_j^{Q'} - x_j^Q)\,. \tag{33}$$

The intersection L of $\overline{QQ'}$ with a hyperplane l that does not belong to the T set satisfies the following relation

$$\xi_q + V\,(\xi_{q'} - \xi_q) = \sum_{j=1}^{N} a_{lj} x_j^L - c_l. \tag{34}$$

If the x_j^L values are now substituted in this expression from (33), after some simple arithmetic manipulations, we arrive at

$$V = \frac{\xi_q - \xi_l}{\sum\limits_{j=1}^{N} a_{lj} x_j^{Q'} - \xi_l - c_l + \xi_q - \xi_{q'}}\,. \tag{35}$$

With V known, L may be determined through (32) and (33). The intersections of the remaining hyperplanes, excluding T and l, will be given by

$$\xi_k^L = \sum_{j=1}^{N} a_{kj} x_j^L - c_k = \sum_{j=1}^{N} a_{kj}\,(x_j^Q + V\,(x_j^{Q'} - c_j^Q)) - c_k \tag{36}$$

where, using (15), it follows that

$$\xi_k^L = (1 - V)\,(\xi_k + c_k) - c_k + V \sum_{j=1}^{N} a_{kj} x_j^Q\,. \tag{37}$$

In a similar way to the previous cases, a feasible point L, common to $T+1$ hyperplanes, is determined for which $\xi_l \geqq \xi_k^L$ for $k = 1, 2, \ldots, 2M$.

Case 4: *Q is a common point of T hyperplanes with* $T \geqq N+1$.
In $(N+1)$-space, the point Q will obviously be the only common point of the T hyperplanes. Two such cases for $N=2$ are illustrated in Figures 5 and 6.

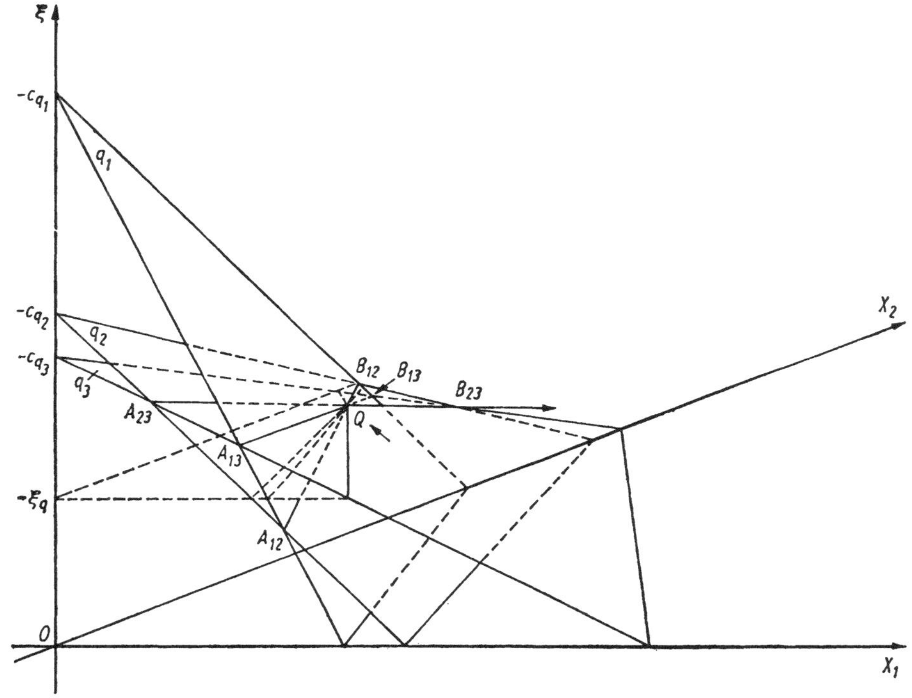

Figure 5. Case of Triple Point with $N=2$ and $c_{q_1}, c_{q_2}, c_{q_3} \leqq 0$

A feasible direction of decreasing ξ may be determined via the intersection of N hyperplanes passing through Q. Each line satisfying this condition must be examined through a procedure identical to that of Case 3.

If none of the candidate line segments forms a feasible direction of decreasing ξ, then the absolute minimum value of ξ, corresponding to Q, has been found and the algorithm terminates successfully.

In the event that such a direction of decreasing ξ exists, then the line segment corresponding to it, according to Case 3, will reveal a new intersection point of T' hyperplanes, where $T' \geqq N+1$, and the above procedure is repeated.

In all four cases outlined above, computation of the intersections is relatively straight forward and may be accomplished rapidly because of the form of the requations describing ξ_k^L and relation (7). The procedure for specifying a feasible point of intersection L is accelerated if the hyperplanes l, for which $\xi_l \geqq 0$, are examined first since those hyperplanes most probably intersect q on the FBS. Moreover, algorithmic convergence is further accelerated by the selection of an optimum initial point. This issue is briefly discussed in the following section.

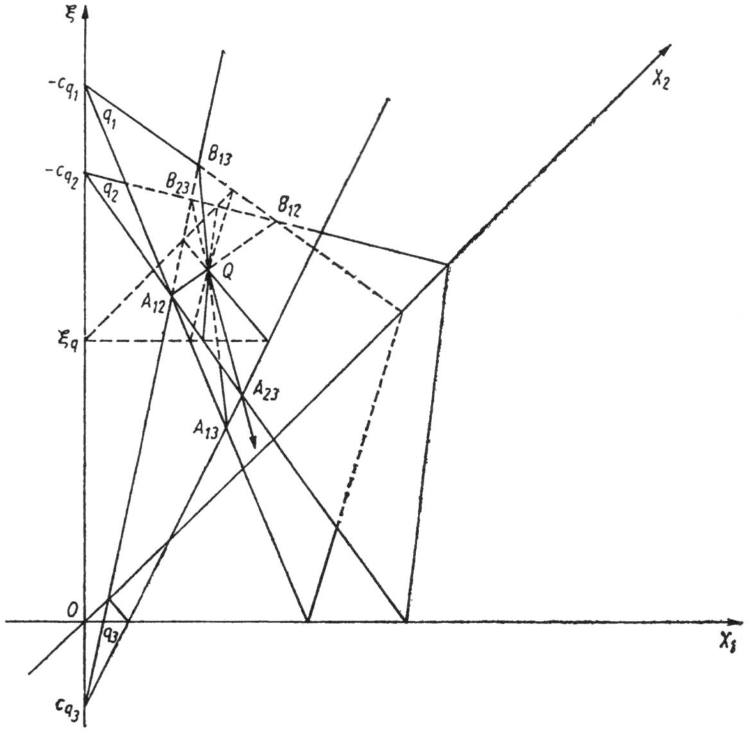

Figure 6. Case of Triple Point with $N=2$ and $c_{q_1}c_{q_2}c_{q_3}>0$

4. Optimum Initial Point

In order for the algorithm to be initiated, some initial feasible point must be located in the search space. The speed of convergence of the algorithm is most certainly influenced by how close this initial point happens to be to the optimum solution.

An optimum initial point may result through the minimization of an L_2 criterion. Such a procedure is shown to lead to the solution of a linear system of algebraic relations with minimum data storage requirements and acceptable solution cycle times.

The objective function to be minimized is of the form:

$$J=\sum_{i=1}^{M}\left(\sum_{j=1}^{N} a_{ij}x_j-c_i\right)^2. \tag{38}$$

Minimization of J may be, of course, accomplished by employing one of the well known nonlinear optimization methods such as that of FLETCHER [11]. In this case, the total number of iterations required for the solution may be not greater than N (quadratic convergence) [12]. None the less, a technique is introduced here which reduces further the computational cost. It is initiated by setting the first derivatives of J to zero, thus resulting in a linear system of N equations.

From (38), we may write

$$\frac{\partial J}{\partial x_k} = 2 \sum_{j=1}^{M} a_{ik} \left(\sum_{j=1}^{N} a_{ij} x_j - c_i \right); \tag{39}$$

and, therefore, the resulting linear system is of the form

$$\sum_{j=1}^{N} h_{ij} x_j = g_i \quad \forall i = 1, 2, \ldots, N \tag{40}$$

where

$$h_{ij} = \sum_{k=1}^{M} a_{kj} a_{ki} \tag{41}$$

and

$$g_i = \sum_{k=1}^{M} c_k a_{ki} . \tag{42}$$

From relations (39) and (41), we observe that the $2h_{ij}$ terms are the elements of the Hessian matrix of J and that $h_{ij} = h_{ji}$. The linear system (40) is, therefore, symmetric about the main diagonal. If $\boldsymbol{H}$ is the array containing h_{ij}, it may be easily inverted using the method of CHOLESKY [13]. That is, it may be written as

$$\boldsymbol{H} = \boldsymbol{U}\boldsymbol{U}^{\mathsf{T}} \tag{43}$$

where $\boldsymbol{U}$ is a lower triangular matrix and $\boldsymbol{U}^{\mathsf{T}}$ its transpose.

Now $\boldsymbol{U}^{-1}$ is given by the relation

$$\boldsymbol{H}^{-1} = (\boldsymbol{U}^{-1})^{\mathsf{T}} \, (\boldsymbol{U}^{-1}) . \tag{44}$$

For the computation of $\boldsymbol{H}^{-1}$, it is only necessary to calculate the inverse of a lower triangular matrix, which, in itself, is also lower triangular and multiply it by its transpose. The derivation of $\boldsymbol{U}^{-1}$ and $(\boldsymbol{U}^{-1})^{\mathsf{T}}$ may be achieved through analytic relations. Thus, the u_{ij}'s may be expressed as

$$u_{ii} = \left(h_{ii} - \sum_{k=1}^{i-1} u_{ik}^2 \right)^{1/2} \tag{45}$$

and

$$u_{ij} = \left(h_{ij} - \sum_{k=1}^{j-1} u_{ik} u_{jk} \right) / u_{jj} \quad j \neq i \tag{46}$$

whereas the u'_{ij} elements of $\bar{U}^{-1}$ are specified as

$$u'_{ii} = 1/u_{ii} \tag{47}$$

and

$$u'_{ij} = -\left(\sum_{k=j}^{i-1} u_{ik} u'_{kj} \right) / u_{ii} \quad \text{for} \quad j \neq i . \tag{48}$$

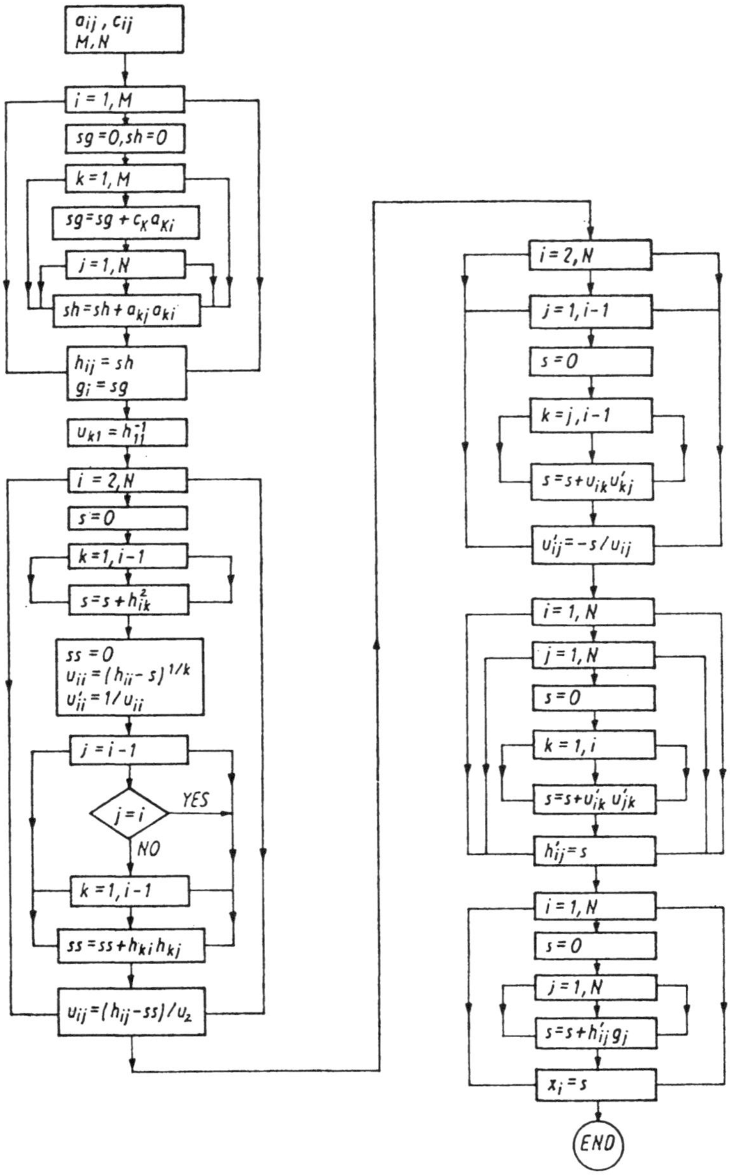

Figure 7. Computer Flow Diagram for the Optimum Initial Point Routine

According to relation (44), the elements of $\boldsymbol{H}^{-1}$, denoted as h'_{ij}, are expressed as functions of u'_{ij} via

$$h'_{ij} = \sum_{k=1}^{i} u'_{ik} u'_{jk} \tag{49}$$

and, finally, the x_i's are equal to

$$x_i = \sum_{j=1}^{N} h'_{ij} g_j \quad \text{for} \quad i=1, 2, \ldots, N\,. \tag{50}$$

The solution of the minimization problem is, therefore, derived analytically through the above algebraic relations. The computational cost is drastically reduced. Furthermore, if the initial a_{ij} coefficients are specified through a suitable subroutine, then the computation of the x_i's may be accomplished through analytic relations with minimum storage requirements. Implementation of the above procedure is straightforward with the corresponding computer flow diagram shown as Figure 7.

5. The Algorithmic Steps

This paragraph summarizes the overall flow diagram of the proposed algorithm. The *procedural steps* for its implementation are as follows:

Step 1: Transform the problem (1) into the form of (3).

Step 2: Select an initial point either through the optimum method described in Paragraph 4 or randomly.

Step 3: Specify, for the point Q, the intersections ξ_k, $k=1, 2, \ldots, 2M$ and the hyperplane q where Q lies.

Step 4: On the basis of the values of ξ_k, $k=1, 2, \ldots, 2M$, specify the number of hyperplanes that have Q as a common point; thus, if Q is a point of:
- only one hyperplane, proceed to Step 5.
- two hyperplanes, go to Step 6.
- T hyperplanes, with $3 \leqq T \leqq N$, go to Step 8.
- T hyperplanes, with $T \geqq N+1$, go to Step 9.

Step 5: According to paragraph 3 for Case 1, draw line $\overline{TQ}$ and specify the intersection point L. Set $Q=L$ and go back to Step 4.

Step 6: If $c_{q_1} c_{q_2} < 0$ or $0 \leqq c_{q_1} \leqq \xi_{q_1}$, proceed to Step 7. Otherwise, consider that Q lies on the hyperplane q_1 (note that $|c_{q_1}| \leqq |c_{q_2}|$) according to paragraph 3, Case 2(a), and go back to Step 5.

Step 7: Specify the x_{A_j} and ξ_{A_j} and the line $\overline{A_j Q}$. Afterwards, according to the procedure of paragraph 3, Case 2(b), specify the intersection point L. Set $Q=L$ and go back to Step 4.

Step 8: Find the second point Q' from the solution of the linear system (31). Then, compute the feasible point of intersection L between the line $\overline{QQ'}$ and a hyperplane. Set $Q=L$ and go back to Step 4.

Step 9: For each line specified from the intersection of N hyperplanes passing through the point Q, check to verify if it constitutes a feasible direction

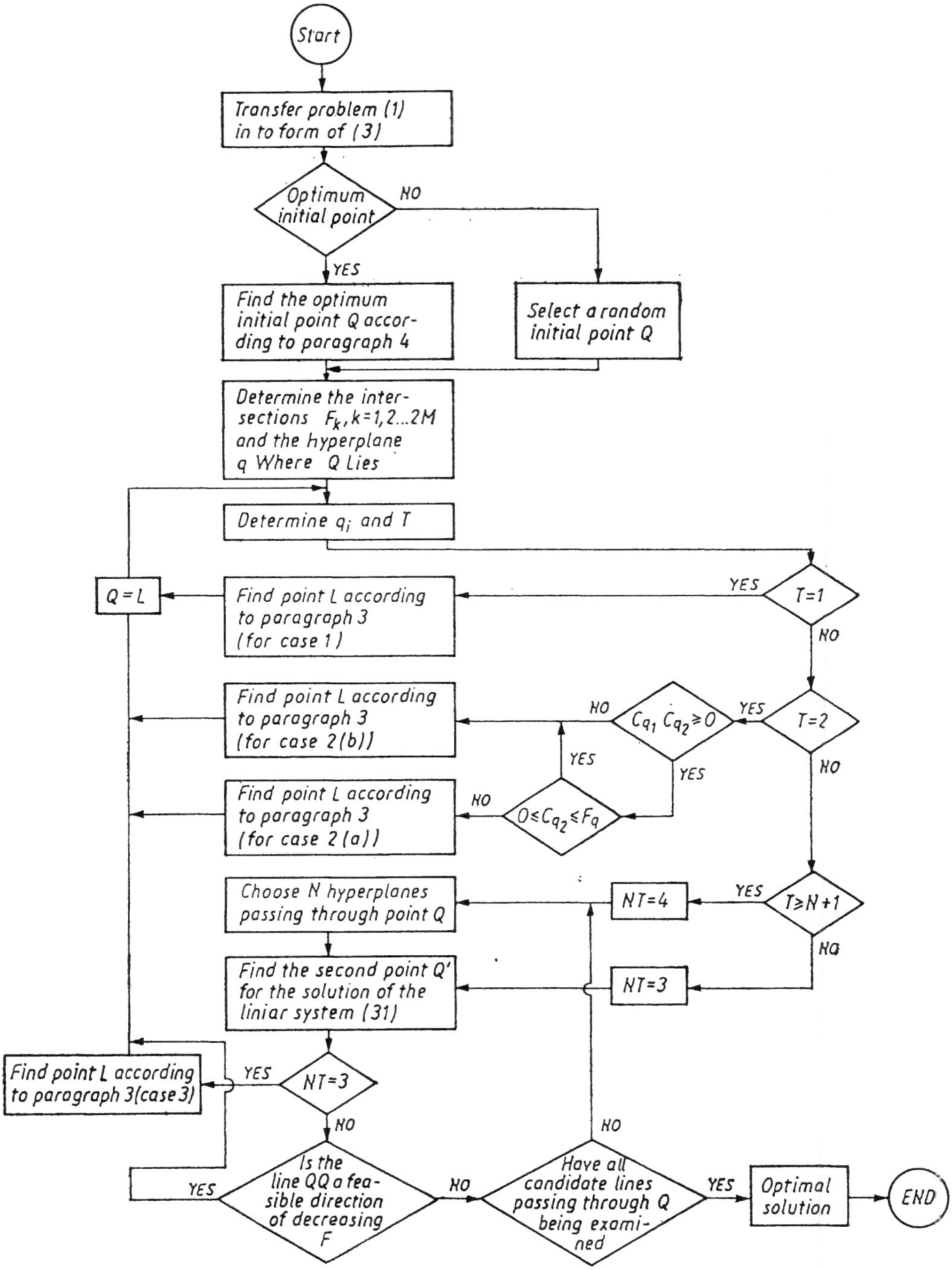

Figure 8. Computer Flow Diagram for the Solution of the Minimization Problem

of decreasing ξ. The procedure for determining the candidate lines is in accordance to Step 8. When the feasible intersection is specified, set $Q=L$ and go back to Step 4. If none of the candidate lines provides a direction of decreasing ξ, then the optimum ξ has been found and the algorithm terminates.

Computation of the intersections ξ_k, $k=1, 2, \dots, 2M$ is accomplished, for the procedural steps outlined above, via the algebraic relations detailed in Paragraph 3. The overall flow chart for the algorithm is depicted diagrammatically as Figure 8.

6. Illustrative Examples

Example 1: In this first example, the following **MINMAXAD** problem is to be solved:

minimize ξ

under the constraints

$$\left|x_1+0x_2+0x_3-1\right| \leqq \xi \tag{51}$$

$$\left|x_1+\frac{1}{6}x_2+\frac{1}{36}x_3-1\right| \leqq \xi$$

$$\left|x_1+\frac{1}{3}x_2+\frac{1}{9}x_3-0.8\right| \leqq \xi$$

$$\left|x_1+\frac{1}{2}x_2+\frac{1}{4}x_3-0.55\right| \leqq \xi$$

$$\left|x_1+\frac{2}{3}x_2+\frac{4}{9}x_3-0.3\right| \leqq \xi$$

$$\left|x_1+\frac{5}{6}x_2+\frac{25}{36}x_3-0.1\right| \leqq \xi$$

where the variables x_1, x_2, x_3 are not sign-constrained. For this problem $N=3$ and $M=6$. If it is solved using the revised Simplex method, the optimum results are:

$$x_1=1.05, \quad x_2=-0.45, \quad x_3=-0.9 \quad \text{and} \quad \xi=0.05\,.$$

On the basis of the proposed algorithm, the optimum initial point Q is specified as

$$x_1=1.042857, \quad x_2=-0.5314286, \quad x_3=-0.7714284 \quad \text{and}$$
$$\xi=0.0671429\,.$$

The point Q is found to lie on the $q=8$ hyperplane with intersections ξ_k, $k==1, 2, \dots, 12$ equal to

$$\begin{array}{ll}
\xi_1= 0.04285714286 & \xi_7=-0.04285714286\\
\xi_2=-0.06714285714 & \xi_8= 0.06714285714\\
\xi_3=-0.02 & \xi_9= 0.02\\
\xi_4= 0.03428571429 & \xi_{10}=-0.03428571429\\
\xi_5= 0.04571428571 & \xi_{11}=-0.04571428571\\
\xi_6=-0.03571428571 & \xi_{12}= 0.03571428571\,.
\end{array}$$

Table 1. Summary of Results for Example 1

ξ	x_1	x_2	x_3	q_i
0.0671428	1.0428571	−0.5314285	−0.7714286	8
0.0556760	1.055676	−0.53796	−0.5276911	1, 8
0.0552254	1.055676	−0.533605	0.7745901	1, 5, 8
0.05	1.05	−0.45	−0.9	1, 4, 5, 8, 12
3.45	4.0	1.5	−3.0	4
2.45	3.0	1.125	−2.25	4, 5
1.16666	1.6	0.6	−1.2	5, 6
0.386046	0.613953	0.2302325	−0.4604651	5, 6, 7
0.2	0.8	0.15	−0.9	5, 6, 7, 8
0.08	1.08	−0.93	−0.18	1, 5, 6, 8
0.05	1.05	−0.45	−0.9	1, 4, 5, 8, 12
31.011111	−5.0	10.0	40.0	6
1.210638	−0.210638	0.4212765	1.685106	6, 7
0.45	0.55	0.0	0.0	6, 7, 8
0.2	0.8	0.15	−0.9	5, 6, 7, 8
0.08	1.08	−0.93	−0.18	1, 5, 6, 8
0.05	1.05	−0.45	−0.9	1, 4, 5, 8, 12

Table 2. Summary of Results for Example 2 (with Optimium Initial Point)

	x_1	x_2	x_3	x_4	x_5	q_i
0.000921963	1.0000351808	0.9977601133	−0.5007063733	0.1765874474	0.0436849972	82
0.000805155	1.000061211	0.997786133	−0.500686854	0.176609905	0.043706278	1, 82
0.000670358	0.999946093	0.997858485	−0.500632579	0.176672351	0.043767121	1, 37, 82
0.0006702	1.0002923095	0.997614332	−0.501217619	0.176916661	0.044005159	1, 16, 37, 82
0.000669862	1.000394188	0.997892998	−0.50182722	0.176638334	0.044513667	1, 16, 17, 37, 82
0.000660628	1.00037239	0.997870559	−0.50186953	0.176680007	0.044577779	1, 16, 17, 37, 46, 82
0.000552608	1.000115755	0.997355952	−0.501211908	0.177292633	0.044176788	1, 15, 16, 37, 46, 82
0.000543921	1.00089025	0.997310344	−0.501143411	0.177340929	0.044185021	1, 15, 37, 46, 82, 69

Table 3. Summary of Results for Example 2 (without Optimum Initial Point Routine)

	x_1	x_2	x_3	x_4	x_5	q_i
1.1321206	0.5	0.5	0.5	0.5	0.5	42
0.8961990	0.264079	0.264079	−0.264079	0.264079	0.264079	70, 42
0.8859179	0.253797	0.295734	0.295734	0.295734	0.295734	42, 69, 70
0.1903512	0.851193	0.833028	−0.459935	0.342173	0.342173	42, 69, 70, 41
0.1184955	0.888315	0.998629	−0.594746	0.176572	0.368007	41, 42, 54, 69, 70
0.0248597	0.976971	0.998629	−0.522972	0.176571	0.113941	41, 42, 53, 69, 70, 54
0.0111759	0.989900	0.998455	−0.512283	0.176746	0.076600	41, 42, 52, 53, 69, 70
0.0065659	0.994360	0.998267	−0.508535	0.176934	0.063821	41, 42, 51, 52, 69, 70
0.0042369	0.996603	0.998063	−0.506515	0.177138	0.057230	41, 42, 50, 51, 69, 70
0.0028313	0.997976	0.997840	−0.505181	0.177361	0.053118	41, 42, 49, 50, 69, 70
0.0018981	0.998905	0.997593	−0.504184	0.177608	0.050257	41, 42, 48, 49, 69, 70
0.0012452	0.999574	0.997320	−0.503370	0.177882	0.048121	6, 41, 42, 48, 69, 70
0.0012194	0.999602	0.997303	−0.503331	0.177898	0.048029	6, 40, 41, 42, 69, 70
0.0009103	0.999919	0.997142	−0.50208	0.177971	0.046792	6, 39, 40, 42, 69, 70
0.0007621	1.000062	0.997091	−0.502524	0.177924	0.046119	5, 6, 39, 42, 69, 70
0.0007249	0.999995	0.997247	−0.501962	0.177694	0.045513	5, 6, 38, 39, 42, 69
0.0006617	1.000165	0.997023	−0.502290	0.177924	0.045613	5, 38, 39, 42, 69, 70
0.0006400	1.000186	0.997014	−0.502233	0.177908	0.045488	5, 16, 38, 42, 69, 70
0.0006095	1.000168	0.997146	−0.501981	0.177445	0.044894	5, 16, 42, 69, 70, 82
0.0005607	1.000111	0.997265	−0.501368	0.177376	0.044337	5, 15, 16, 42, 69, 82
0.0005439	1.000089	0.997310	−0.501143	0.177347	0.044135	5, 15, 37, 42, 69, 82

This problem was solved using three different initial points. The solution procedure is shown in Table 1. For each step of the algorithm, the location of Q, the values of the x_j variables, the corresponding value of ξ, as well as those hyperplanes that Q is a common point of, are specified. The algorithmic steps leading to the optimum solution are clearly illustrated. The dependence of the number of iterations on the location of the initial search point selected is also depicted in the table.

Example 2: The proposed algorithm is now applied to a problem of larger dimensions. The functions $G(x)=e^x-x^2$ is to be approximated by the polynomial

$$f(x)=a_1+a_2x+a_3x^2+a_4x^3+a_5x^4 . \tag{52}$$

The sampling interval is $[-1, 1]$ and 41 sample points are to be used in approximating $G(x)$.

The revised Simplex method provides the following solution

$$x_1= 1.000089025 \qquad x_4=0.177346929$$
$$x_2= 0.997310344 \qquad x_5=0.044135021$$
$$x_3=-0.501143411 \quad \text{and} \quad \xi=0.000543921 .$$

The optimum initial point Q is found to be equal to

$$x_1= 1.0000351808 \qquad x_4=0.1765874464$$
$$x_2= 0.9977601133 \qquad x_5=0.0436843972$$
$$x_3=-0.5007063733 \quad \text{and} \quad \xi=0.00092106 ,$$

and Q lies on the $q=82$ hyperplanes.

Table 2 shows the detailed results derived from the application of the proposed algorithm with an optimum point used to initiate the scheme. Table 3 exhibits similar results but now a random initial point equal to $x_i=0.5$ for $i=1, 2, ..., 5$ was used.

Comparing the results from the two cases, we may note the significant effect of choosing the optimum initial point on the speed of convergence of the algorithm. Moreover, the results from Table 3 demonstrate the effective convergence properties of the algorithm.

7. Comparative Results and Conclusions

The minimax approximation problem is well known in the technical literature. Its solution has been approached thus far primarily through Linear Programming (LP) techniques. LP methods usually entail high computational costs, large memory requirements, and roundoff errors that cannot be neglecteed.

The proposed algorithm introduces a new approximation technique to the problem. It is based upon the full utilization of the formulation of the problem constraints and the feasible region. It specifies feasible directions of decreasing values of the objective function resulting always in a global minimum. A definite advantage of the algorithm relates to the arbitrary choice of an initial point. But speed of convergence may be substantially improved by a judicial choice of an optimum initial point through a linear method.

The algorithm was applied to a number of simulations using SBASIC on a microcomputer. In order to arrive at some valid comparative results, the same computational environment was used to host the revised Simplex algorithm. Thus, the simulation time using RSA was approximately 17.6 sec. for the problem of Example 1. Implementation of the new algorithm reduced the solution time to 4.9, 11.2, and 7.3 sec., respectively, for the three cases studied. Also, for Example 2, the revised Simplex method required 210.9 sec. for a solution, whereas the proposed algorithm reduced the solution time to 116.7 sec. as depicted in Table 2.

References

[1] CHARALAMBOUS, C.: The performance of an algorithm for minimax design of two-dimensional linear phase FIR digital filters. IEEE Trans. Circuits Syst., CAS-**32** (1985) 1016–1028.

[2] MCCLELLAN, J. H.; T. W. PARKS; L. R. ROBINER: A computer program for designing optimum FIR linear phase digital filters. IEEE Trans. Audio Electroacoustics, AU-**21** (1973) 506–625.

[3] MEINARDUS, G.: Approximation of Functions: Theory and Numerical Methods. Springer Tracts in Natural Philosophy, **13** (1967).

[4] ARTHANARI, T. S.; Y. DODGE: Mathematical Programming in Statistics. John Wiley & Sons, 1981.

[5] System/360 Scientific Subroutine Package. IBM Application Program, Version III, Programmer's Manual, prog. number 360A-CM-03X, 5th ed., August 1970.

[6] PAPADEMETRIOU, C.; K. STEIGLITZ: Combinatorial Optimization: Algorithms and Complexity. Prentice-Hall, 1982.

[7] CONN, A. R.: Nonlinear programming, exact penalty functions and projection techniques for non-smooth functions. Numerical Optimization, SIAM, (1985), pp. 3–25.

[8] FLETCHER, R.: A model algorithm for composite nondifferentiable optimization problems. Mathematical Programming Study, **17** (1982) 67–76.

[9] MURRAY, W.; M. OVERTON: A projected Lagrangian algorithm for nonlinear minimax optimization. SISSC, **1** (1980) 345–370.

[10] WATSON, G. A.: Approximation Theory and Numerical Methods, John Wiley, Crichester and New York, 1980.

[11] FLETCHER, R.: A new approach to variable metric algorithms. Computer J., **13** (1970) 317–322.

[12] GOTTFRIED, B. S.; J. WEISMAN: Introduction to Optimization Theory. Prentice-Hall, Inc., Englewood Cliffs, NJ, 1973.

[13] FORSYTHE, G.; C. B. MOLER: Computer Solution of Linear Algebraic Systems. Prentice-Hall, Inc., NJ, 1967.

Received August 1987, revised January 1988

N. Papamarkos
Democritus University of Thrace
School of Engineering
Xanthi, Greece

G. Vachtsevanos
Georgia Institute of Technology
School of Electrical Engineering
Atlanta, Georgia 30332-0250
U.S.A.

optimization 19 (1988) 6, 861–873

Thickness Optimization Problem

G. ANICULAESEI
University of Iași, Institute of Mathematics

Summary: An optimal control problem governed by a variational inequality of elliptic type is considered. Necessary optimality conditions are obtained and, in some special cases, the optimal control is determined.

AMS 1980 Subject Classification:

Primary: 49 B 34; Secondary: 49 A 29

Key words: Optimal control, optimality conditions, optimal design.

0. Introduction

The paper is devoted to the study of an optimal control problem which simulates the optimization of the energy associated with the thickness u of a membrane covering a domain Ω with boundary $\partial\Omega$ and with vertical deformation $y(x)$ due to the pressure f. Many problems of this type arise in structural engineering (shells, in particular), and can be formulated as an optimal control problem with the control u described by coefficients of a PDE, see BANICHUK [2] for example. The problem we are concerned with has been studies earlier by CEA and MALANOWSKI [5] and PIRONNEAU [10], but our result and the method we use differ from theirs.

Our contribution consists in that we establish a set of necessary conditions for optimality in a form which enables us to say more about the optimal control in the case of good domains.

The paper is organized according to the following plan. Sections 1 and 2 contain the statement of the problem, and the existence for the control problem respectively.

Following the lines of BARBU [3] in treating the control problems governed by variational inequalities we derive in Section 3 the optimality conditions and then by using the concept of "viscosity solution" introduced by CRANDALL and LIONS [6] we give, in Section 4, a formula for the computation of optimal control which is made explicit in two special cases.

As regards optimality conditions we mention that comparable results may be obtained using the concept of directional differentiability of solutions to variational inequalities developed by F. MIGNOT [9].

1. Statement of the Problem

Let $\Omega \subset \boldsymbol{R}^2$ be a bounded domain with the boundary $\partial\Omega$ sufficiently smooth.

The following usual notation will be used in the sequel: $V = H_0^1(\Omega)$, $H = L^2(\Omega)$, $V' = H^{-1}(\Omega)$. The norms and scalar products in V, H, V' will be denoted by $\|.\|$ and $(,)$ respectively. The notation $|.|$ will be used to designate the modulus in $\boldsymbol{R}$.

We introduce the set U of control functions:

$$U = \{u \in W^{1,\infty}(\Omega) \mid 0 < a \leqq u(x) \leqq b, \mid \nabla u \mid \leqq \varrho \text{ a.e. in } \Omega, u = u_0 \text{ on } \partial\Omega\}$$

where $u_0 \in C(\overline{\Omega})$ satisfies $a \leqq u_0 \leqq b$ in Ω, and $a < u_0 < b$ in $\partial\Omega$; and ϱ is a given positive number.

Consider the control problem

(P) *Minimize*

$$J(u) = \int_\Omega u(x) \, |\nabla y(x)|^2 \, dx \tag{1.1}$$

on the set of all pairs $(y, u) \in V \times U$ which satisfy the variational inequality

$$\int_\Omega u \nabla y \nabla (y - z) \, dx \leqq \int_\Omega (f, y - z) \, dx , \quad \forall z \in K \tag{1.2}$$

where

$$K = \{y \in H_0^1(\Omega) \mid y \geqq 0, \text{ a.e. } x \in \Omega\}$$

and $f \in L^2(\Omega)$ is a given function.

Let us observe that (1.2) can also be written in the form

$$\begin{cases} -\nabla (u \nabla y) + \beta(y) \ni f & \text{a.e. } x \in \Omega \\ y = 0 & x \in \partial\Omega \end{cases} \tag{1.2'}$$

where β is the subgradient of the indicator function corresponding to the closed, convex set K, i.e.

$$\beta(r) = \begin{cases} 0 & \text{if } r > 0 \\ \boldsymbol{R}^- & \text{if } r = 0 \\ \varnothing & \text{otherwise.} \end{cases} \tag{1.3}$$

For fixed $u \in U$ define the operator $A(u): V \to V'$ by

$$(A(u) \, y, z) = \int_\Omega u \nabla y \nabla z \, dx .$$

This is linear, bounded, and coercive, hence by Corollary 2.1 [3, p. 40] the equation (1.2)' has a unique solution.

2. Existence for the Problem (P)

Definition 2.1: A pair $(y^*, u^*) \in V \times U$ for which the infimum in Problem (P) is attained is called an *optimal pair* and the corresponding control u^* – *optimal control.*

Proposition 2.1: *The Problem* (P) *has at least one optimal pair.*

Proof: Firstly let us note that $J(u)$ is lower semicontinuous in the weak star topology of $L^\infty(\Omega)$.
To prove this let u, $u_n \in U$ be such that:

$$u_n \to u \quad \text{weakly star in} \quad L^\infty(\Omega)\,, \tag{2.1}$$

and y (respectively y_n) the solutions corresponding to u (u_n) in (1.2)'.
Using the definition of J we obtain

$$\begin{aligned} J(u_n) - J(u) &= \int_\Omega u_n\, |\nabla y_n|^2 \, dx - \int_\Omega u\, |\nabla y|^2 \, dx \\ &= \int_\Omega u_n\, |\nabla\ (y_n - y)|^2 \, dx - \int_\Omega (u_n - u)\, |\nabla y|^2 \, dx \\ &\quad + 2 \int_\Omega (u_n \nabla y_n \nabla y - u\, |\nabla y|^2)\, dx\,. \end{aligned}$$

Now, (y, u) and (y_n, u_n) satisfy the equations

$$\begin{cases} -\nabla(u \nabla y) + \beta(y) \ni f & \text{a.e. in } \Omega \\ -\nabla(u_n \nabla y_n) + \beta(y_n) \ni f & \text{a.e. in } \Omega \\ y,\ y_n = 0 & \text{in } \partial\Omega\,. \end{cases}$$

Multiplying the above equations by y it follows that

$$\int_\Omega (u_n \nabla y_n \nabla y - u\, |\nabla y|^2 \, dx = \int_\Omega (\beta(y) - \beta(y_n))\, y \, dx\,.$$

Thus we have

$$\begin{aligned} J(u_n) - J(u) &= \int_\Omega u_n\, |\nabla\ (y_n - y)|^2 \, dx - \int_\Omega (u_n - u)\, |\nabla y|^2 \, dx \\ &\quad + 2 \int_\Omega (\beta(y) - \beta(y_n))\, y \, dx\,. \end{aligned}$$

Inasmuch as y_n, $y \in K$, the definition of β implies

$$(\beta(y) - \beta(y_n))\, y \geqq 0\,.$$

Then, letting $n \to \infty$ in (2.2) yields

$$\liminf_{n\to\infty} (J(u_n) - J(u)) \geqq \liminf_{n\to\infty} \int_\Omega u_n\, |\nabla\ (y_n - y)|^2 \, dx \geqq 0\,.$$

Thus

$$\liminf J(u_n) \geqq J(u)\,. \tag{2.3}$$

Let now $u_n \in U$ be a minimizing sequence for J, that is

$$\lim J(u_n) = \inf_{u \in U} J(u)\,.$$

Since U is bounded, closed and convex in $L^\infty(\Omega)$ it follows that it is compact in the weak star topology of this space. Hence there exists on $u^* \in U$ and a subsequence $\{u_{nk}\}$ of $\{u_n\}$ such that

$$u_{nk} \to u^* \quad \text{weakly star in} \quad L^\infty(\Omega)\,.$$

Having in mind (2.3) it follows

$$J(u^*) \leqq \lim\inf J(u_{n_k}) = \lim J(u_n) = \inf_{u \in U} J(u) ,$$

hence (y^*, u^*), (y^* corresponding to u^* in (1.2')) is an optimal pair for Problem (P). ■

3. Necessary Conditions of Optimality

The approximating control process. Given $\varepsilon > 0$ and u^* an optimal control for Problem (P), consider the approximating problem as follows.

(P_ε) *Minimize*

$$\int_\Omega \left(u \; |\nabla y|^2 + \frac{1}{2} \; (u - u^*)^2\right) \mathrm{d}x \tag{3.1}$$

on the set of pairs $(y, u) \in V \times U$ satisfying:

$$\begin{cases} -\nabla (u \nabla y) + \beta^\varepsilon(y) = f & \text{a.e.} \quad x \in \Omega \\ y = 0 & \quad x \in \partial\Omega \end{cases} \tag{3.2}$$

where β^ε is a smooth approximation of β [3, p. 84] given by

$$\beta_\varepsilon(r) = -\frac{r^-}{\varepsilon}$$

$$\beta^\varepsilon(r) = \int_{-\infty}^{\infty} (\beta_\varepsilon \; (r - \varepsilon^2 \vartheta) - \beta_\varepsilon \; (-\varepsilon^2 \vartheta)) \; \varrho(\vartheta) \; \mathrm{d}\vartheta , \tag{3.3}$$

ϱ being a positive mollifier: $\varrho(\vartheta) = \varrho(-\vartheta)$, $\varrho(\vartheta) \geqq 0$, $\int_{-1}^{1} \varrho(\vartheta) \, \mathrm{d}\vartheta = 1$, supp $\varrho \subset [-1,1]$ $\varrho \in C^\infty(\boldsymbol{R})$.

Doing the calculations in (3.3) yields

$$\beta^\varepsilon(r) = \varepsilon^{-1} \int_{\varepsilon^{-2}r}^{1} (r - \varepsilon^2 \vartheta) \varrho(\vartheta) \; \mathrm{d}\vartheta + \varepsilon \int_0^1 \vartheta \varrho(\vartheta) \; \mathrm{d}\vartheta , \tag{3.4}$$

and therefore

$$|\beta^\varepsilon(r) - \beta_\varepsilon(r)| \leqq 2\varepsilon , \quad \forall r \in \boldsymbol{R} . \tag{3.5}$$

Let $(y_\varepsilon, u_\varepsilon)$ be an optimal pair for (P_ε) whose existence follows in the same way as for (P).

On the other hand since β^ε is differentiable, the mapping $u \to y_\varepsilon(u)$ is GÂTEAUX differentiable from U to V and for all $v \in L^\infty(\Omega)$, $z_\varepsilon = \nabla y_\varepsilon(u_\varepsilon) v \in V$, satisfies the equation

$$\begin{cases} \nabla (u_\varepsilon \nabla z_\varepsilon) + \nabla (v \nabla y_\varepsilon) - \dot{\beta}^\varepsilon(y_\varepsilon) z_\varepsilon = 0 & \text{in} \quad \Omega \\ z_\varepsilon = 0 & \text{in} \quad \partial\Omega . \end{cases} \tag{3.6}$$

Now, we can state the optimality conditions for (P_ε).

Proposition 3.1: *There exists a $p_\varepsilon \in V$ which together with y_ε and u_ε satisfies the* *optimality system*

$$\begin{cases} -\nabla(u_\varepsilon \nabla y_\varepsilon) + \beta^\varepsilon(y_\varepsilon) = f & \text{in } \Omega \\ y_\varepsilon = 0 & \text{in } \partial\Omega \end{cases} \tag{3.7}$$

$$\begin{cases} -\nabla(u_\varepsilon \nabla p_\varepsilon) + \dot\beta^\varepsilon(y_\varepsilon)\, p_\varepsilon = -2\nabla(u_\varepsilon \nabla y_\varepsilon) & \text{in } \Omega \\ p_\varepsilon = 0 & \text{in } \partial\Omega \end{cases} \tag{3.8}$$

$$\nabla y_\varepsilon \nabla(p_\varepsilon - y_\varepsilon) \in \partial I_U(u_\varepsilon) + u_\varepsilon - u^* \quad \text{a.e. in } \Omega. \tag{3.9}$$

Proof: We denote by I_U the indicator function of U i.e.

$$I_U(u) = \begin{cases} 0 & \text{if } u \in U \\ +\infty & \text{if } u \notin U . \end{cases}$$

Let $v \in L^\infty(\Omega)$ and $\lambda > 0$ be sufficiently small, such that

$$\int_\Omega \left((u_\varepsilon + \lambda v)\, |\nabla y_\varepsilon^\lambda|^2 + I_U\,(u_\varepsilon + \lambda v) + \frac{1}{2}\,(u_\varepsilon + \lambda v - u^*)^2 \right) dx$$
$$- \int_\Omega \left(u_\varepsilon |\nabla y_\varepsilon|^2 + \frac{1}{2}\,(u_\varepsilon - u^*)^2 \right) dx \geqq 0 ,$$

where y_ε^λ is the solution of the equation (1.2)′ corresponding to $(u_\varepsilon + \lambda v)$. Dividing by λ, and letting $\lambda \to 0$ in the above relation we get

$$\int_\Omega (2u_\varepsilon \nabla y_\varepsilon \nabla z_\varepsilon + v|\nabla y_\varepsilon|^2 + I_U'(u_\varepsilon, v) + (u_\varepsilon - u^*)v)\, dx \geqq 0 \tag{3.10}$$

where

$$I_U'(u_\varepsilon, v) = \lim_{\lambda \to 0} (I_U\,(u_\varepsilon + \lambda v) - I_U(u_\varepsilon))/\lambda .$$

Multiplying (3.6) by p_ε and (3.8) by z_ε, we obtain after subtraction

$$2 \int_\Omega u_\varepsilon \nabla y_\varepsilon \nabla z_\varepsilon dx = - \int_\Omega v \nabla y_\varepsilon \nabla p_\varepsilon dx \tag{3.11}$$

which along with (3.10) implies (3.9). ■

In order to pass to the limit in (3.7)–(3.9) we need some estimates of the quantities appearing in these approximating equations. Multiplying (3.7) by y_ε, integrating on Ω and taking into account that $\beta^\varepsilon(r)\, r \geqq 0$ we get

$$\int_\Omega u_\varepsilon |\nabla y_\varepsilon|^2\, fx \leqq \int_\Omega (f, y_\varepsilon)\, dx$$

which implies that $\{\nabla y_\varepsilon\}$ is bounded in $L^2(\Omega)$, hence $\{y_\varepsilon\}$ is bounded in $H_0^1(\Omega)$.

In the same way, multiplying (3.7) by $\beta^\varepsilon(y_\varepsilon)$ and using that $\nabla \beta^\varepsilon(y^\varepsilon) \nabla y_\varepsilon \geqq 0$ we obtain, after integrating on Ω, that $\{\beta^\varepsilon(y_\varepsilon)\}$ is bounded in $L^2(\Omega)$.

Now (3.7) implies that $\{u_\varepsilon \nabla y_\varepsilon\}$ is bounded in $W^{1,2}(\Omega)$. The boundedness of $\{p_\varepsilon\}$ in $H_0^1(\Omega)$ is obtained after multiplication of (3.8) by p_ε and having in mind that $\dot\beta^\varepsilon(r) \geqq 0$. Coming back to (3.8) we conclude that $\{\nabla(u_\varepsilon \nabla p_\varepsilon)\}$ is bounded in $L^1(\Omega)$.

Summarizing, up to now we have obtained the following information about the sequences appearing in (3.7)–(3.9).

$$\{y_\varepsilon\} \quad \text{is bounded in} \quad H_0^1(\Omega) , \tag{3.12}$$

$$\{\beta^\varepsilon(y_\varepsilon)\} \quad \text{is bounded in} \quad L^2(\Omega) , \tag{3.13}$$

$$\{u_\varepsilon \nabla y_\varepsilon\} \quad \text{is bounded in} \quad W^{1,2}(\Omega) , \tag{3.14}$$

$$\{p_\varepsilon\} \quad \text{is bounded in} \quad H_0^1(\Omega) , \tag{3.15}$$

$$\{\dot\beta^\varepsilon(y_\varepsilon)\, p_\varepsilon\} \quad \text{is bounded in} \quad L^1(\Omega) , \tag{3.16}$$

$$\{u_\varepsilon \nabla p_\varepsilon\} \quad \text{is bounded in} \quad W^{1,1}(\Omega) . \tag{3.17}$$

Since $\Omega \subset \boldsymbol{R}^2$, the following two inclusions are compact

$$W^{1,1}(\Omega) \subset L^p(\Omega) \quad \text{for} \quad 1 \leqq p < 2 , \tag{3.18}$$

and

$$W^{1,2}(\Omega) \subset L^q(\Omega) \quad \text{for} \quad 1 \leqq q \leq \infty \tag{3.19}$$

Now we prove

Proposition 3.2: *The following two convergences hold*

$$u_\varepsilon \to u^* \quad \textit{strongly in } L^2(\Omega) ,$$
$$y_\varepsilon \to y^* \quad \textit{weakly in } H_0^1(\Omega) \textit{ and strongly in } L^2(\Omega) .$$

Remark: In fact, the above convergences hold on some subsequences of $\{\varepsilon\}$, but for simplicity we have not changed the notation.

Proof of Proposition 3.2: We begin with the equation

$$\begin{cases} -\nabla(u^* \nabla y) + \beta^\varepsilon(y) = f & \text{a.e. in } \Omega \\ y = 0 & \text{in } \partial\Omega \end{cases} \tag{3.20}$$

and denote the solution of this equation by y_ε^*.
From (3.12) it follows that there exists an $\tilde y \in H_0^1(\Omega)$ such that

$$y_\varepsilon^* \to \tilde y \quad \text{weakly in } H_0^1(\Omega) \text{ and strongly in } L^2(\Omega) . \tag{3.21}$$

We shall prove that $\tilde y = y^*$. From (3.13) we get

$$\beta^\varepsilon(y_\varepsilon^*) \to \zeta \quad \text{weakly in} \quad L^2(\Omega)$$

which together with (3.21) and (3.5) implies $\zeta \in \beta(\tilde y)$.
Multiplying (3.20) by $\varphi \in H_0^1(\Omega)$ and passing to the limit with $\varepsilon \to 0$ we obtain

$$\int_\Omega (u^* \nabla \tilde y \nabla \varphi + \beta(\tilde y)\, \varphi)\, \mathrm{d}x = \int_\Omega f \varphi \mathrm{d}x , \quad \forall \varphi \in H_0^1(\Omega)$$

and having in mind that (1.2′) has an unique solution it follows that $\tilde y = y^*$. This together with (3.14) (where we take $u^* \nabla y_\varepsilon^*$ instead of $u_\varepsilon \nabla y_\varepsilon$) and (3.21), leads to

$$\int_\Omega u^* |\nabla y_\varepsilon^*|^2 \,\mathrm{d}x \to \int_\Omega u^* |\nabla y^*|^2 \,\mathrm{d}x . \tag{3.22}$$

Coming back to (3.7) we note that there exists $\bar{\bar y} \in H_0^1(\Omega)$ such that $y_\varepsilon \to \bar{\bar y}$ weakly in $H_0^1(\Omega)$ and stronly in $L^2(\Omega)$ and since U is a compact subset of $L^2(\Omega)$ there exists an $\bar{u} \in U$ such that

$$u_\varepsilon \to \bar{u} \quad \text{strongly in } L^2(\Omega) .$$

Now, multiplying (3.7) by $\varphi \in H_0^1(\Omega)$ and having in mind (3.5), we obtain upon passing to the limit

$$-\nabla (\bar{u} \nabla \bar{\bar{y}}) + \beta(\bar{\bar{y}}) \ni f ,$$

and since the equation (1.2) has an unique solution it follows that $\bar{\bar{y}} = \bar{y}$ (the solution corresponding to $\bar{u}$). Passing to the limit with $\varepsilon \to 0$ in the inequality

$$\int_\Omega \left(u_\varepsilon |\nabla y_\varepsilon|^2 + \frac{1}{2} (u_\varepsilon - u^*)^2 \right) dx \leqq \int_\Omega u^* \; |\nabla y^*|^2 \; dx$$

and having in mind (3.22) we obtain

$$\int_\Omega \left(u |\nabla y|^2 + \frac{1}{2} (u - u^*)^2 \right) dx \leqq \int_\Omega u^* |\nabla y^*|^2 \; dx ,$$

which implies that $\bar{u} = u^*$ and $\bar{y} = y^*$ thereby completing the proof. ■

Remark 3.1: From the proof of Proposition 3.2 it incidentally follows that the Problem (P_ε) approximates the Problem (P) as well as the passage to the limit in (3.7).

Let us now analyze the equation (3.8). From (3.15) it follows that there exists $p^* \in H_0^1(\Omega)$ such that $p_\varepsilon \to p^*$ weakly in $H_0^1(\Omega)$ and strongly in $L^2(\Omega)$.

Then (3.16) implies that $\{\dot{\beta}^\varepsilon(y_\varepsilon) \, p_\varepsilon\}$ is weakly star compact in $(L^\infty(\Omega))'$, hence, there exists a measure ν such that on a generalized subsequence $\{\lambda\}$ of $\{\varepsilon\}$,

$$\dot{\beta}_\lambda(y_\lambda) \, p_\lambda \to \nu \quad \text{weakly star in } (L^\infty(\Omega))' . \tag{3.23}$$

Now, multiplying (3.8) by $\varphi \in C_0^\infty(\Omega)$ and passing to the limit with $\varepsilon \to 0$ we get

$$\nabla (u^* \nabla p^*) - \nu = 2 \nabla (u^* \nabla y^*) \text{ in } D'(\Omega) . \tag{3.24}$$

In what follows we try to give a more exact interpretation of the last equality.

For this we shall use the same method as that used in the proof of the Theorem 3.3 [3, p. 83].

Following [3] let $\zeta_\varepsilon, \lambda_\varepsilon : \Omega \to \boldsymbol{R}$ be measurable functions given by

$$\zeta_\varepsilon(x) = \begin{cases} 0 & \text{if} \quad |y_\varepsilon(x)| > \varepsilon^2 \\ 1 & \text{if} \quad |y_\varepsilon(x)| \leqq \varepsilon^2 , \end{cases}$$

and

$$\lambda_\varepsilon(x) = \begin{cases} 0 & \text{if} \quad y_\varepsilon(x) > -\varepsilon^2 \\ 1 & \text{if} \quad y_\varepsilon(x) \leqq -\varepsilon^2 . \end{cases}$$

Since

$$\dot{\beta}_\varepsilon(r) = \varepsilon^{-1} \int_{\varepsilon^{-2} r}^{1} \varrho(\vartheta) \, d\vartheta , \quad \text{for all} \quad r \in \boldsymbol{R} \tag{3.25}$$

this together with (3.4) implies

$$|y_\varepsilon(x) \, \dot{\beta}^\varepsilon(y_\varepsilon(x)) \, p_\varepsilon(x) - p_\varepsilon(x) \, \beta^\varepsilon(y_\varepsilon(x))| \leqq \varepsilon |p_\varepsilon(x)| \quad \text{a.e. } x \in \Omega . \tag{3.26}$$

Now from (3.4), (3.25) and the definition of ζ_ε and λ_ε it follows

$$|p_\varepsilon\beta^\varepsilon(y_\varepsilon)| \leqq \varepsilon|p_\varepsilon\dot{\beta}^\varepsilon(y_\varepsilon)|\ (\varepsilon^{-1}|y_\varepsilon|\zeta_\varepsilon+\varepsilon^{-1}|y_\varepsilon|\lambda_\varepsilon)+2\varepsilon|p_\varepsilon| \quad \text{a.e. in} \quad \Omega\,. \tag{3.27}$$

But (3.4) and (3.13) imply that

$$\beta^\varepsilon(y_\varepsilon)\ \lambda_\varepsilon = \varepsilon^{-1}y_\varepsilon\lambda_\varepsilon + C\varepsilon\lambda_\varepsilon\,,$$

C being a constant independent of ε, is bounded in $L^2(\Omega)$.

Now since $\varepsilon^{-1}|y_\varepsilon|\,\zeta_\varepsilon \leqq \varepsilon$ a.e. in Ω, it follows by (3.15), (3.16), (3.25) and (3.27) that a subsequence $\varepsilon_n \to 0$

$$p_{\varepsilon_n}(x)\ \beta^{\varepsilon_n}(y_{\varepsilon_n}(x)) \to 0 \quad \text{a.e.} \quad x\in\Omega\,. \tag{3.28}$$

On the other hand, (3.13) and (3.15) together with (3.7) yield

$$p_{\varepsilon_n}\beta^{\varepsilon_n}(y_{\varepsilon_n}) \to p^*\ (f+\nabla(u^*\nabla y^*)) \quad \text{weakly in} \quad L^1(\Omega)$$

which, together with (3.28) and the Egorov Theorem, gives

$$p^*\ (f+\nabla(u^*\nabla y^*))=0 \quad \text{a.e.} \quad x\in\Omega\,,$$

consequently

$$p_{\varepsilon_n}\beta^{\varepsilon_n}(y_{\varepsilon_n}) \to p^*\ (f+\nabla(u^*\nabla y^*)) \quad \text{strongly in} \quad L^1(\Omega)\,,$$

and from (3.26) we obtain

$$y_{\varepsilon_n}\dot{\beta}^{\varepsilon_n}(y_{\varepsilon_n})\ p_{\varepsilon_n} \to 0 \quad \text{strongly in} \quad L^1(\Omega)\,. \tag{3.29}$$

Now according to the Egorov Theorem, for every $\eta>0$ there exists a subset Ω_η of Ω such that $m(\Omega\backslash\Omega_\eta)<\eta$, and $y_{\varepsilon_n}\to y^*$ uniformly in Ω_η.

Thus by (3.23) and (3.29) $y^*\nu=0$ in Ω_η, that is

$$(y^*\nu_a)\ (\Phi)+\nu_s(y^*\varnothing)=0\,, \quad \text{for all} \quad \Phi\in L^\infty(\Omega) \quad \text{with} \quad \operatorname{supp}\Phi\subset\Omega_\eta \tag{3.30}$$

where ν_a and ν_s are the absolutely continuos part and the singular part of the measure ν, respectively. But, since ν_s is singular there exists an increasing sequence Ω_k such that $m(\Omega\backslash\Omega_k)\leqq k^{-1}$ and $\nu_s=0$ on $L^\infty(\Omega_k)$.

This along with (3.30) would imply that $(y^*\nu_a)\,(\Phi)=0$, for all $\Phi\in L^\infty(\Omega)$, with support $\Phi\in\Omega_k\cap\Omega_\eta$. From this, after letting k tend to infinity and then η tend to zero we get $y^*\nu_a=0$ a.e. in Ω. Together with (3.24) the latter yields

$$(\nabla(u^*\nabla p^*))_a = 2(\nabla(u^*\nabla y^*))_a \quad \text{a. e. in} \quad [y^*>0]\,,$$

where the subscript a stands for "absolutely continuous".

Passing now to (3.9), let us observe that (3.14), (3.19) and Proposition 3.2 imply that $\{\nabla y_\varepsilon\}$ is compact in $L^q(\Omega)$ for all $q\geqq 1$, while (3.17), (3.18) yield that $\{\nabla p_\varepsilon\}$ is compact in $L^p(\Omega)$, $1\leqq p<2$.

These assertions imply

$$\nabla y_\varepsilon\nabla\ (p_\varepsilon-y_\varepsilon) \to \nabla y^*\nabla\ (p^*-y^*) \quad \text{strongly in} \quad L^1(\Omega)\,.$$

Multiplying (3.9) by $u_\varepsilon-v$, $v\in L^\infty(\Omega)$, and letting ε tend to zero, we get

$$\int_\Omega \nabla y^*\nabla\ (p^*-y^*)\,(u^*-v)\ \mathrm{d}x \geqq 0\,, \quad \text{for all} \quad v\in U$$

which implies

$$\nabla y^* \nabla\ (p^* - y^*) \in \partial I_U(u^*) \quad \text{a.e.} \quad x \in \Omega\ .$$

Summarizing, we have obtained the following necessary contions of optimality for the Problem (P).

Theorem 3.1: *If* (y^*, u^*) *is an optimal pair for the Problem* (P) *then there exists the dual arc* $p^* \in H_0^1(\Omega)$ *and a measure* $\nu \in (L^\infty(\Omega))'$ *which along with* y^* *and* u^* *satisfies the system*

$$-\nabla (u^* \nabla y^*) + \beta(y^*) \ni f \qquad in \quad \Omega \tag{3.31}$$

$$\nabla (u^* \nabla p^*) - \nu = 2 \nabla (u^* \nabla y^*) \qquad in \quad D'(\Omega) \tag{3.32}$$

$$-(\nabla (u^* \nabla p^*))_a = 2f \qquad a.e.\ in \quad [y^* > 0] \tag{3.33}$$

$$p^*\ (f + \nabla (u^* \nabla y^*)) = 0 \qquad a.e. \quad x \in \Omega \tag{3.34}$$

$$\nabla y^* \nabla\ (p^* - y^*) \in \partial I_U(u^*) \qquad a.e. \quad x \in \Omega\ . \tag{3.35}$$

4. Calculating the Optimal Control

In the following we shall show that in some cases, we may obtain sufficient information about the optimal control u^* from Theorem 3.1. To this aim we need the structure of ∂I_U (the subdifferential of the function I_U in the sense of ROCKAFELLAR [11]). This is given by

Lemma 4.1: *Assume that there is a* $g \in W^{1,\infty}(\Omega)$ *such that* $g = u_0$ *in* $\partial\Omega$ *and* $\|g\|_{L^\infty(\Omega)} < \varrho$. *Then for any* $u \in U$, $w \in \partial I_U(u)$ *if and only if* w *has the form*

$$w = -\operatorname{div} \vartheta + \eta \quad in \quad \Omega \tag{4.1}$$

where $\quad \vartheta \in (L^2(\Omega))^2$,

$$\vartheta(x) = \begin{cases} 0 \quad a.e.\ in \quad \{x \in \Omega \mid |\nabla u(x)| < \varrho\} \\ \lambda(x) \nabla u(x) \quad a.e.\ in \quad \{x \in \Omega \mid |\nabla u(x)| = \varrho\} \\ where \quad \lambda \in L^2(\Omega), \quad \lambda(x) \geqq 0 \quad a.e. \quad x \in \Omega \end{cases} \tag{4.2}$$

and

$$\eta \in L^2(\Omega)$$

$$\eta(x) = \begin{cases} \ \ 0 \quad a.e.\ in \quad \{x \in \Omega \mid a < u(x) < b\} \\ \leqq 0 \quad a.e.\ in \quad \{x \in \Omega \mid u(x) = a\} \\ \geqq 0 \quad a.e.\ in \quad \{x \in \Omega \mid u(x) = b\}\ . \end{cases} \tag{4.3}$$

Proof: See [4, Theorem 4.1]. ■

In the sequel, we assume that the following condition will be in effect

$$f \in L^2(\Omega), \quad f > 0 \quad \text{a.e. in} \quad \Omega\ . \tag{*}$$

Using the strong maximum principle for elliptic equations [7, Theorem 8.19] it follows that $y^*(x) > 0$ for all $x \in \Omega$.

At the same time (3.31) and (3.33) imply

$$\nabla y^*(x)\nabla\big(p^*(x)-y^*(x)\big)>0 \quad \text{a.e.} \quad x\in\Omega\,. \tag{4.4}$$

From (4.2) we observe that

$$\nabla u^*(x)=\varrho\,\frac{\vartheta(x)}{|\vartheta(x)|} \quad \text{a.e. in} \quad \{x\in\Omega \mid \vartheta(x)\neq 0\}\,. \tag{4.5}$$

On the other hand from (3.35), (4.1) and (4.3) we get

$$|\nabla u^*(x)|=\varrho \quad \text{a.e. in} \quad Q=\{x\in\Omega \mid u^*(x)<b\}\,. \tag{4.6}$$

In other words, $u^*\in W^{1,\infty}(\Omega)$ is a weak solution in Q of the eikonal equation

$$|\nabla u|=\varrho\,. \tag{4.7}$$

Recall [8, p. 48) that in general equation (4.7) with Dirichlet condition on the boundary has an infinite number of weak solutions (i.e., solutions $u\in W^{1,\infty}(Q)$ which satisfy a.e. the equation). Crandall and Lions [6] have introduced the notion of "viscosity solution" for a class of equtions which includes (4.7) and which solves, among others, the problem of uniqueness. For convenience let us recall one of their definitions. We set

$$D^+u(x_0)=\{p\in \boldsymbol{R}^2 \mid \limsup_{x\to x_0}\big(u(x)-u(x_0)-p(x-x_0)\big)|x-x_0|^{-1}\leqq 0\}\,,$$

$$D^-u(x_0)=\{p\in \boldsymbol{R}^2 \mid \limsup_{x\to x_0}\big(u(x)-u(x_0)-p(x-x_0)\big)|x-x_0|^{-1}\geqq 0\}\,.$$

Then $u\in C(Q)$ is a viscosity solution for (4.7) if

$$|p|\leqq\varrho \quad \text{for any} \quad x_0\in Q,\quad p\in D^+u(x_0)\,,$$

$$|p|\geqq\varrho \quad \text{for any} \quad x_0\in Q,\quad p\in D^-u(x_0)\,.$$

From this definition we remark that if $u\in W^{1,\infty}(Q)$ is a viscosity solution for (4.7) then it is also a weak solution for it.

A function $u\in C(Q)$ is said to be *semiconcave* if for any $\delta>0$ there exists a $C_\delta>0$ such that $u(x)-C_\delta|x|^2$ is concave on every convex subset of

$$Q_\delta=\left\{x\in Q \mid |x|<\frac{1}{\delta},\ \text{dist}\,(x,\partial Q)>\delta\right\}.$$

A function $u\in L^1_{\text{loc}}(Q)$ is said to be *semi-super harmonic* if for any $\delta>0$ there exists a $C_\delta>0$ such that

$$\Delta u\leqq C_\delta \quad \text{in} \quad D'(Q)\,.$$

Notice that if u is semiconcave, then it is also semi-super harmonic.

Returning to (4.7) we introduce the function L as follows [8, p. 137].

$$L(x,y)=\inf\{\varrho T \mid \exists\zeta\in\text{Lip}\,[0,T],\ \zeta(0)=x,\ \zeta(T)=y;\ \zeta(t)\in Q,$$
$$\forall t\in[0,T];\ |\zeta'(t)|\leqq 1 \quad \text{a.e.} \quad t\in[0,T]\}$$

if $(x,y)\in Q\times Q$.

If $x\in Q$, $y\in\partial Q$ we define $L(x,y)$ by

$$L(x,y)=L(y,x)=\liminf_{\substack{z\to y\\ z\in Q}} L(x,z)\leqq\infty\,.$$

Finally if $x, y \in \partial Q$ we define $L(x, y)$ by

$$L(x, y) = L(y, x) = \liminf_{\substack{t \to x \\ z \to y \\ (t,z) \in Q}} L(t, z) \leqq \infty .$$

Also, let $\Phi : \partial Q \to \mathbf{R}$ be defined by

$$\Phi(y) = \begin{cases} u_0(y) & \text{if} \quad y \in \partial Q \cap \partial \Omega \\ b & \text{if} \quad y \in \partial Q \setminus \partial \Omega . \end{cases} \tag{4.8}$$

Now we can prove the following

Theorem 4.2: *The function u^* is a viscosity solution of the eikonal equation (4.7) in Q. Moreover, u^* is semiconcave in any connected component $\tilde{Q}$ of Q, and it is given by*

$$u^*(x) = \inf_{y \in \partial \tilde{Q}} \{\Phi(y) + L(x, y)\} , \quad \forall x \in \tilde{Q} . \tag{4.9}$$

Proof: In virtue of Proposition 5.2 in [8, p. 137] it suffices to show that u^* is the maximum element of the set

$$S = \{v \in W^{1,\infty}(\tilde{Q}) | \ |\nabla v| \leqq \varrho \text{ in } \tilde{Q}, \ v \leqq u^* \text{ in } \partial \tilde{Q}\} .$$

But from (4.5) for any $v \in S$ we have

$$- \int_{\tilde{Q}} \operatorname{div} \vartheta \nabla \ (v - u^*)^+ \, dx = \int_{\tilde{Q}} \vartheta \nabla \ (v - u^*)^+ \, dx \leqq 0 .$$

Now, since (4.4) implies $-\operatorname{div} \vartheta > 0$ a.e. in $\tilde{Q}$, the above relation gives $(v - u^*)^+ = 0$ a.e. in $\tilde{Q}$, hence

$$u^*(x) \geqq v(x) \quad \text{a.e. } x \in \tilde{Q}, \quad \text{and for any} \quad v \in S$$

so we conclude. ■

Remark 4.1: By uniqueness of semi-super harmonic viscosity solutions [8, p. 82] we conclude that u^* is the unique semi-super harmonic viscosity solution to (4.7) with boundary condition (4.8). Notice that this does not imply uniqueness of u^* because $\tilde{Q}$ depends on u^*.

However, in case Ω has a particular form we can determine u^*, which implies its uniqueness.

Notice that (4.6) and (4.9) imply $\partial \tilde{Q} \cap \partial \Omega \neq \Phi$ for any connected component $\tilde{Q}$ of Q.

Indeed if $\partial \tilde{Q} \cap \partial \Omega \neq \emptyset$ we get

$$b > u^*(x) = b + \inf_{y \in \partial \tilde{Q}} L(x, y) , \quad \forall x \in \tilde{Q}$$

which is a contradiction, since $L(x, y) \geqq :0$.

This implies that if $\partial \Omega$ is connected, then Q is connected, too.

If we assume $u_0(x) = u_0 = \text{const.}$ and $\partial \Omega$ is connected, then Theorem 4.2 yields

$$u^*(x) = u_0 + \inf_{y \in \partial \Omega} L(x, y) , \quad \forall x \in Q . \tag{4.10}$$

In the following we give two examples in which the last formula becomes explicit.

Example 4.1: Let

$$\Omega=\{(x, y)\in \boldsymbol{R}\times \boldsymbol{R} \mid |x|\leqq \alpha,\ |y|\leqq \beta,\ \alpha\geqq \beta>0\}\ .$$

In this case

$$\inf_{(z,t)\in\partial\Omega} L((x, y), (z, t))=\varrho d(x, y)$$

where

$$d(x, y)=\operatorname{dist}((x, y), \partial\Omega)$$

and from (4.10) we find

$$u^*((x, y)) = \begin{cases} u_0+\varrho d(x, y), & \text{for any } (x, y)\in\Omega \text{ for which } d((x, y))<\dfrac{b-u_0}{\varrho}, \\ b & \text{otherwise.} \end{cases} \tag{4.11}$$

We make the following notations

$$\tilde{\Omega}=\{(x, y)\in\Omega \mid |y|\leqq|\alpha-\beta-x \operatorname{sgn} x|\}$$

$$\omega_1=\left\{(x, y)\in\Omega \mid |x|>\alpha-\frac{b-u_0}{\varrho}\right\}$$

$$\omega_2=\left\{(x, y)\in\Omega \mid |y|>\beta-\frac{b-u_0}{\varrho}\right\}.$$

From simple geometric considerations, we easily deduce that the control u^* has the following form

$$u^*(P)=\begin{cases} u_0+\varrho(\alpha-x \operatorname{sgn} x) & \text{if } P\in\tilde{\Omega}\cap\omega_1 \\ u_0+\varrho\ (\beta-y \operatorname{sgn} y) & \text{if } P\in(\Omega\setminus\tilde{\Omega})\cap\omega_2 \\ b & \text{otherwise}\ , \end{cases}$$

where $P=P(x, y)\in\Omega$.

Example 4.2: Let

$$\Omega=\{x \mid \|x\|\leqq r\}\ ,$$

where $x\in\boldsymbol{R}^2$ and $\|x\|$ denotes its usual euclidean norm. In this case (4.10) yields

$$u^*(x)=\begin{cases} u_0+\varrho\ (r-\|x\|) & \text{if } \|x\|>r-\dfrac{b-u_0}{\varrho} \\ b & \text{if } \|x\|\leqq r-\dfrac{b-u_0}{\varrho}\ . \end{cases}$$

Acknowledgement

I would like to thank Professor V. Barbu for introducing the topic to me, and for his stimulating comments and suggestions. I wish also to thank the referees for their helpful comments.

References

[1] ADAMS, R.; SOBOLEV spaces. Academic Press, New York 1975.
[2] BANICHUK, N.: Problems and Methods of Optimal Structural Design. Plenum Press, New York and London 1983.
[3] BARBU, V.: Optimal Control of Variational Inequalities. Pitman, London 1984.
[4] BARBU, V.; A. FRIEDMAN: Optimal design of domains with free boundary problems. SIAM J. Contrcl and Opt. (to appear).
[5] CEA, J.; K. MALANOWSKI: An example of max-min problems in partial differential equations. SIAM J. Control and Opt., **8** (1970) 3, 305–316.
[6] CRANDALL, M. G.; L. C. EVANS; P. L. LIONS: Some properties of viscosity solutions of HAMILTON-JACOBI equations. Trans. Amer. Math. Soc., **282** (1984) 487–502.
[7] GILBARG, D.; N. S. TRUDINGER: Elliptic Partial Differential Equations of Second Order. Springer, Berlin 1983.
[8] LIONS, P. L.: Generalized solutions of HAMILTON-JACOBI Equations. Pitman, London 1982.
[9] MIGNOT, F.: Contrôle dans les inéquations variationnelles elliptiques. J. Functional Analysis, **22** (1976) 130–185.
[10] PIRONNEAU, O.: Optimal Shape Design for Elliptic Systems. Springer-Verlag, New York 1983.
[11] ROCKAFELLAR, R. T.: Convex Analysis. Princeton University Press, Princeton, N. J., 1970.

Received April 1987, revised December 1987

G. Aniculaesei
University of Iaşi
Institute of Mathematics
6600 Iaşi
Romania

optimization 19 (1988) 6, 874

Book Review

HOFMANN, B.: **Regularization for Applied Inverse and Ill-Posed Problems.** Teubner Texte zur Mathematik. 85. Teubner Leipzig, 1987, 196 S., M 19,–, ISBN 3-322-00341-8.

This monograph is concerned with the solution of linear and non-linear inverse and ill-posed problems. After a brief introduction into the general problem, in Chapter 2 the phenomenon of illposedness is discussed by means of some classical applied inverse problems: linear integral equations of the first kind, computerized tomography, backward heat equation, boundary control, inverse eigenvalue problems.

In Chapter 3 the connections between optimal control and parameter identification problems are elucidated by means of a general optimization approach. Chapter 4 discusses methods to solve deterministic problems by regularization techniques: TIKHONOV-regularization, multi-parameter-regularization, adaptive regularization. The subsequent Chapter 5 brings a treatment of the regularization in discrete stochastic inverse problems. The final Chapter 6 gives some hints for the actual numerical calculation of solutions.

The book is well-written and easy to read. It gives a nice introductory treatment into inverse and ill-posed problems, and it seems to be very suitable also for non-mathematicians. The author follows consequently his intention to make the reader familiar with the fundamental idea of regularization. Special attention is given to the in practical applications particularly important discrete problems, but the connection to their underlying continuous origins is never lost. In this respect, the author has found a good compromize.

J. SPREKELS

optimization 19 (1988) 6, 875–888

An Enumeration Algorithm for Combinatorial Problems of the Reliability Analysis of Binary Coherent Systems

F. Zettwitz

Ingenieurhochschule Mittweida, Sektion Informationselektronik

Summary: In this paper the problem of the generation of all elements of a system of sets is investigated and a backtrack algorithm (A1) is given solving this problem. The algorithm is applied to combinatorial problems of reliability:
– A t-graph is interpreted as a binary coherent system $\underline{S}$ and the system function of $\underline{S}$ is represented as the sum of Boolean orthogonal products. This approach is a simple and efficient alternative to a method suggested by Satyanarayana and N. Hagstrom [9].
– A directed p-graph is also interpreted as a binary coherent system $\underline{S}$ and both the minimal cuts of $\underline{S}$ and the cocycles of G are determined.

AMS 1980 Subject Classification:

Primary: 05 A 15; Secondary: 05 C 30, 62 N 05

Key words: Enumeration, combinatorics, reliability, graph theory.

1. Introduction

When investigating the reliability of networks enumeration problems occur, such as the generation of all spanning trees of a graph or the generation of all cuts separating two given nodes of a connected graph. These problems can be considered as special cases of a more general problem.

Assuming the set $N=\{1, 2, ..., n\}$ is given. All subsets of N with a given well-defined property form a system of sets E. The enumeration problem consists in the complete and unique generation of all elements of E. In this paper a backtrack algorithm for solving these problems is described which is a generalization of an algorithm from [11].

2. Fundamentals

The system of sets E is characterized by the following conditions.

Condition 1: *E contains the intersection X^0 of its elements. X^0 is known.*

Condition 2: *There is at least one sequence for two elements at a time $X', X'' \in E$ with $X' \subset X''$*

$$X_1, X_2, ..., X_j, \quad (j>1) \tag{1}$$

of elements from E for which

$$X' = X_1 \subset X_2 \subset \ldots \subset X_j = X''$$

and $|X_{i+1} \backslash X_i| = 1$ *for* $i = 1, 2, \ldots, j-1$ *are valid.* In the following $A + B$ are written instead of $A \cup B$, iff A and B are disjoint sets.

Let $X \in E$ and $Y \subseteqq N \backslash X$, then the following definitions are valid:

Definition 1: An element $v \in N \backslash X$ is called the *candidate* of X, if $X + \{v\} \in E$ is valid. Note: For two neighbours of the sequende (1) X_i and X_{i+1} there is a candidate v of X_i, so that $X_i + \{v\} = X_{i+1}$ holds.

Definition 2: The set $C(X) = \{v \mid v \in N \backslash X \text{ and } X + \{v\} \in E\}$ is called the *set of candidates of X.*

Definition 3: The set $C(X, Y) = C(X) \backslash Y$ is called the *set of candidates of X reduced by Y.*

Definition 4: The system of sets $E(X, Y) = \{Z \in E \mid Z \supseteqq X \text{ and } Z \cap Y = \varnothing\}$ is called a *(X, Y)-system of E.* Note: From Condition 1 and Definition 4 follows $E = E(X^0, \varnothing)$.

Lemma 1: *Let* $X \in E$ *and* $Y \subseteqq N \backslash X$. $E(X, Y) = \{X\}$ *is valid iff* $C(X, Y)$ *is empty.*

Proof: Let $E(X, Y) = \{X\}$ and let $C(X, Y)$ be not empty. Then there is at least one candidate v in $C(X, Y)$ with $X + \{v\} \in E(X, Y)$. Now let $C(X, Y)$ be empty with one element $X'' \supset X$ of E contained in $E(X, Y)$. Then, according to Condition 2 there is a sequence (1) with $X' = X$ the elements of which are contained in $E(X, Y)$. For the second element X_2 of this sequence $X + \{v\} = X_2$ with $v \notin Y$ holds, i.e. $C(X, Y)$ would be not empty. ∎

Theorem 1 (Partition Theorem): *Let* $X \in E$, $Y \subseteqq N \backslash X$ *and* $C(X, Y) = \{v_1, v_2, \ldots, v_r\}$. *Then there is a partition of* $E(X, Y)$ *into pairwise disjoint subsystems*

$$E(X, Y) = \{X\} + E(X + \{v_1\}, Y\} + E(X + \{v_2\}, Y + \{v_1\}) + \ldots$$
$$+ E(X + \{v_r\}, Y + \{v_1, v_2, \ldots, v_{r+1}\}) . \tag{2}$$

Proof: These sets from $E(X, Y)$ containing v_1 (not containing v_1) form the system of sets $E(X + \{v_1\}, Y)$ ($E(X, Y + \{v_1\})$). Thus, $E(X, Y) = E(X + \{v_1\}, Y) + E(X, Y + \{v_1\})$ holds. Decomposing $E(X, Y + \{v_1\})$ in respect to v_2, you will get $E(X, Y + \{v_1\}) = E(X + \{v_2\}, Y + \{v_1\}) + E(X, Y + \{v_1, v_2\})$ and so $E(X, Y) = E(X + \{v_1\}, Y) + E(X + \{v_2\}, Y + \{v_1\}) + E(X, Y + \{v_1, v_2\})$. By continuing decomposition up to v_r (2) is formed whereby $E(X, Y + \{v_1, \ldots, v_r\}) = \{X\}$ holds. ∎

3. Partition Procedure

Every $X \in E$ is assigned to a well-determined sequence of pairs of sets. Thereby the following notations are used: If $X \in E$ with $|X \backslash X^0| = k$, then instead of X we write more precicely $X = X^k$.

Let $X = X^j$ with $X^j \in E$, $j > 0$. If, according to Lemma 1, $E = E(X^0, \emptyset)$ is decomposed into non-empty disjoint subsystems, X^j will be contained in exactly one subsystem $E(X^1, Y_1)$. Thus, the first two pairs

$$(X^0, Y_0) \tag{3}$$

$(Y_0 = \emptyset)$ and (X^1, Y_1) of the sequence are determined. After k partition steps $(0 < k < j)$ let $X^j \in E(X^k, Y_k)$ with $C(X^k, Y_k) = \{v_1^k, \ldots, v_r^k\}$. If this (X^k, Y_k)-subsystem is decomposed and $X^j \in E\ (X^{k+1}, Y_{k+1})$ holds then from Theorem 1 follows

$$X^{k+1} = X^k + \{v_{i_k}^k\} \tag{4}$$

with

$$v_{i_k}^k \in C(X^k, Y_k) \tag{5}$$

$$Y_{k+1} = Y_k + \{v_1^k, \ldots, v_{i_k-1}^k\} \quad \text{for} \quad i_k > 1 \tag{6}$$

or

$$Y_{k+1} = Y_k \qquad \text{for} \quad i_k = 1\,. \tag{7}$$

So the pair (X^{k+1}, Y_{k+1}) is formed. A sequence of partition steps up to $j = k$ generates all pairs of the sequence.

Lemma 2: *Any $X^j \in E$, $(j > 0)$ is assigned the sequence*

$$(X^0, Y_0), \quad (X^1, Y_1), \ldots, (X^j, Y_j) \tag{8}$$

by means of the partition procedure, with

$$X^0 \subset X^1 \subset \ldots \subset X^j\,, \tag{9}$$

$X^k \in E$ $(k = 0, 1, \ldots, j)$. *For the sets of the sequence* (8) *the relations* (3), (5), (6) *and* (7), $k = 0, 1, \ldots, j-1$ *hold. For* $j = 0$ (3) *holds.*

By this it is possible to generate all elements of E by the following algorithm.

4. Algorithm (A1) for the Enumeration of all Elements X of the System of Sets E

```
k:=0; Y0:=∅; kand (X^0, Y0); output (X^0, Y0);
while k>=0 do
begin if C(X^k, Yk)≠∅ then
      begin
          choose (X^k, Yk, vk);
          X^(k+1):=X^k+{vk};
          Y(k+1):=Yk;
             k:=k+1;
          output (X^k, Yk);
          kand (X^k, Yk)
      end
    else
```

```
        begin
            k:=k-1;
            if  k>=0  then  Y_k:=Y_k+{v_k};
        end;
end; {while}
```

Notes: – kand (X, Y) is a procedure generating $C(X, Y)$
– choose selects and deletes the first element v from $C(X, Y)$.

Geometric interpretation: By (A1) a directed tree $H(E)$ is generated, the nodes of which are well-defined assigned to the pairs in Lemma 2. The root is assigned to the pair (X^0, Y_0) with $Y_0=\emptyset$. k is the level of $H(E)$, $k \geqq 0$.

Lemma 3: *If the pair (X^k, Y_k) with $C(X^k, Y_k)=\{v_1^k, \ldots, v_r^k\}$ is assigned to a node P^k of the tree $H(E)$ there are exactly r adjacent nodes $P_1^{k+1}, \ldots, P_r^{k+1}$ which the pairs of sets*

$$(X^k+\{v_1^k\}, Y_k)$$
$$(X^k+\{v_2^k\}, Y_k+\{v_1^k\})$$
$$(X^k+\{v_r^k\}, Y_k+\{v_1^k, \ldots, v_{r-1}^k\})$$

belong to.

This Lemma follows from Theorem 1. It is the geometric representation of this Theorem.

5. Applications in the Theory of Reliability

The applications being considered in the following can be classified in the reliability analysis of binary coherent systems. That's why at first some fundamental concepts of this theory are introduced, also for the limitation in respect to analog terms used in the graph theory ([2], [3], [5], [7]).

5.1. Binary Coherent Systems

A binary system and its elements can only acquire the two states of either operating or failing. Assuming the state of the binary system $\underline{S}$ depends only on the states of its elements e_i $(i=1, 2, \ldots, n)$. A component of the Boolean state vector $\underline{z}=(z_1, z_2, \ldots, z_n)$ is assigned to an element e_i. z_i is either equal to 1 if e_i is operating or equal to 0 if e_i has failed.

The state of the binary system $\underline{S}$ is described by the Boolean system function $\varphi(\underline{z})$. $\varphi(\underline{z})$ is equal to 1 iff $\underline{S}$ is operating. The binary system is referred to coherent if $\varphi(\underline{z})$ is an increasing (not decreasing) function in every single variable. A state vector $\underline{z}$ is referred to a path vector if $\varphi(\underline{z})=1$ holds. The set $\{i \mid z_i=1\}$ is then called a *path*. A path is called *minimal* if it has no strict subset, which, in turn, is a path. If $\varphi(\underline{z})=0$ is valid, $\underline{z}$ is referred to a cut vector and the set $\{i \mid z_i=0\}$ is referred to a cut. A cut is called *minimal* if there is no strict subset which, in turn, is cut. $\underline{S}$ is operating (has failed) iff at least the elements of a minimal path (of a minimal cut) are operating (have failed). Each element of $\underline{S}$ is relevant.

5.2. Problems of the Reachability

Assume a t-graph $G=(V,N)$ with the source s is given. A t-graph in respect to s is a digraph with the property that each of its arcs belongs to a rooted spanning tree of G with the root s [3], [7].

The graph G can be interpreted as a binary coherent system $\underline{S}$, whose elements e_i $(i=1, 2, \ldots, n)$ correspond to the arcs of G, which are either operating or have failed. Assuming the nodes of G are absolutely reliable, $\underline{S}$ is referred to operating if starting with s all the other nodes in G can be reached via operating arcs. This exactly happens if the arcs of at least one spanning tree being a rooted tree with the root s are operating. Thus, the set of arcs of these spanning trees are the minimal paths of $\underline{S}$.

Generation of all rooted trees of G: If X is the set of all arcs of a rooted tree of G with the root s, then its set of nodes V_X is determined by s and the terminal nodes of its arcs. The tree is denoted by $T(X)=(V_X, X)$. The root s is considered as a (degenerated) tree $T(X^0)=(\{s\}, \emptyset)$ with $X^0=\emptyset$.

The system of sets E consists of the sets X just introduced. E satisfies condition 1 : $X^0 \in E$.

Candidate: If an arc v from $N \setminus X$ of a node of the tree $T(X)$ leads to an other node j not being in V_X, then v is called a candidate of X since $X+\{v\} \in E$ holds.

Conclusion: *If the conditions $X' \subset X''$ and $X' \neq X''$ are valid for $X', X'' \in E$, then $T(X')$ is a subgraph of $T(X'')$. Let $D=X'' \setminus X'$. For $X_1=X'$ a candidate v_1 of X_1 is chosen from D, $X_2=X_1+\{v_1\}$ is formed, and v_1 is removed from D.*

In the following, beginning with X_2 this procedure is being continued until $D=\emptyset$ holds. It is impossible, that no candidate is found for a set of the sequence (1) with $D \neq \emptyset$ being valid. In this case $T(X'')$ would not be a rooted tree of G. Consequently, Condition 2 is satisfied. When applying the adjacency matrix of G it is rather simple to find the candidates of X. ∎

So (A1) generates the sets of arcs of the trees $T(X)$. If the *spanning trees* with the root s are wanted the output procedure is changed in such a way that only the pairs (X, Y) with $|X|=|V|-1$ are put out. Furthermore, (A1) solves the problem of representing the system function $\varphi(\underline{z})$ from $\underline{S}$ as the sum of orthogonal Boolean products: Two Boolean functions f and g are referred to as orthogonal iff $f \cdot g \equiv 0$ holds. If $f_1, f_2, \ldots, f_n$ are pairwise orthogonal then, instead of $f_1 \cup f_2 \cup \ldots \cup f_n$ you write the sum $f_1+f_2+\ldots+f_n$.

Considering all pairs (X, Y) for which $T(X)$ is a spanning tree and the Boolean product

$$\bigwedge_{i \in X} z_i \bigwedge_{j \in Y} \bar{z}_j \tag{10}$$

is assigned to each pair, then these products are pairwise orthogonal and their sum is the representation of $\varphi(\underset{\sim}{z})$. If $Y=\emptyset$, then the second product must be replaced by 1. The proof of this conclusion and its precision is the object of the following explanations.

A *cut* of $\underset{\sim}{S}$ is a subset of the set N of G containing at least one arc of every spanning tree with the root s. If these arcs fail no path of $\underset{\sim}{S}$ will be operating: $\underset{\sim}{S}$ has failed.

Let $X\in E$. If $T(X)$ is not a spanning tree the set of candidates $C(X)$ will not be empty. The arcs of $C(X)$ have the property that their initial nodes are contained in V_X and their terminal nodes in $V\setminus V_X$. Since the nodes in $V\setminus V_X$, excluding the arcs of $C(X)$, can not be reached from s, $C(X)$ is a cut of $\underset{\sim}{S}$. From Definition 3 follows $C(X)\subseteqq C(X, Y)+Y$. If $C(X, Y)=\{v_1, \dots, v_r\}$ and if Y is not a cut of $\underset{\sim}{S}$ then $Y+\{v_1, \dots, v_r\}$ is a cut which is not necessarily minimal. Resulting from this, there is a smallest index m for which the cut is

$$Y+\{v_1, v_2, \dots, v_m\} \quad (m\leqq r)\,. \tag{11}$$

Partition into sums of orthogonal products

The following *assignment* is introduced:

Let (X, Y) be a pair generated by algorithm (A1) (in what follows we consider exclusively such types of pairs). If $\varphi(\underset{\sim}{z})$ is the system function of $\underset{\sim}{S}$ the well-determined Boolean function

$$F(\underset{\sim}{z}, X, Y)=\bigwedge_{i\in X} z_i\cdot\bigwedge_{j\in Y}\bar{z}_j\cdot\varphi(\underset{\sim}{z}') \tag{12}$$

is assigned to the pair.

For $\underset{\sim}{z}'$ holds: $z_i'=1$ for $i\in X$, if $X\neq\emptyset$, $z_j'=0$ for $j\in Y$, if $Y\neq\emptyset$ and $z_k'=z_k$ for $k\in$ $\in N\setminus X\setminus Y$ if the latter set is not empty. If X or Y are empty the corresponding product in (12) must be replaced by 1.

Properties of the assignment: $F(\underset{\sim}{z}, \emptyset, \emptyset)=\varphi(\underset{\sim}{z})$ holds. If X is the set of arcs of a spanning tree with the root s then $\varphi(\underset{\sim}{z}')=1$ holds and

$$F(\underset{\sim}{z}, X, Y)=\bigwedge_{i\in X} z_i\cdot\bigwedge_{j\in Y}\bar{z}_j \tag{13}$$

is valid.

If Y is any cut of $\underset{\sim}{S}$ then $F(\underset{\sim}{z}, X, Y)=0$. For $\varphi(\underset{\sim}{z})$

$$\varphi(\underset{\sim}{z})=F(\underset{\sim}{z}, \emptyset, \emptyset)=F(\underset{\sim}{z}, \{i\}, \emptyset)+F(\underset{\sim}{z}, \emptyset, \{i\}) \tag{14}$$

holds with $i\in N$.

If a tree $T(X)$ which is not a spanning tree belongs to (X, Y) and if Y is not a cut then $C(X, Y)$ is not empty and $\varphi(\underset{\sim}{z}')$ not constant. Then for every $v\in C(X, Y)$ we get

$$F(\underset{\sim}{z}, X, Y)=F(\underset{\sim}{z}, X+\{v\}, Y)+F(\underset{\sim}{z}, X, Y+\{v\})\,. \tag{15}$$

In order to prove (15) we consider the product $\bigwedge_{i\in X} z_i\cdot\bigwedge_{j\in Y}\bar{z}_j\cdot\varphi(\underset{\sim}{z}')$ and apply Formula (14) for $i=v$ to $\varphi(\underset{\sim}{z}')$.

Theorem 2: *Assuming that (X, Y) are given with Y not being a cut of $\underline{S}$ and $T(X)$ not being a spanning tree. If $C(X, Y)=\{v_1, v_2, \ldots, v_r\}$ and if m is the smallest index for which $Y+\{v_1, v_2, \ldots, v_m\}$ is a cut $(m \leqq r)$ then $F(\underline{z}, X, Y)$ may be represented as a sum of pairwise orthogonal Boolean products:*

$$F(\underline{z}, X, Y)=F(\underline{z}, X+\{v_1\}, Y)+F(\underline{z}, X+\{v_2\}, Y+\{v_1\})+\ldots \\ +F(\underline{z}, X+\{v_m\}, Y+\{v_1, v_2, \ldots, v_{m-1}\}) . \quad (16)$$

Proof: $F(\underline{z}, X, Y)=F(\underline{z}, X+\{v_1\}, Y)+F(\underline{z}, X, Y+\{v_1\})$ holds. If the second summand is decomposed we will obtain

$$F(\underline{z}, X, Y+\{v_1\})=F(\underline{z}, X+\{v_2\}, Y+\{v_1\})+F(\underline{z}, X, Y+\{v_1, v_2\})$$

and consequently
$F(\underline{z}, X, Y)=F(\underline{z}, X+\{v_1\}, Y)+F(\underline{z}, X+\{v_2\}, Y+\{v_1\})+F(\underline{z}, X, Y+\{v_1, v_2\})$.
With further decompositions the last summand $F(\underline{z}, X, Y+\{v_1, \ldots, v_m\})$ is equal to 0. ■

Representation of the system function: Furthermore the tree $H(E)$ generated by (A1) is considered. k is the level of $H(E)$ and indicates the cardinality of the sets X^k. The Boolean function $F(\underline{z}, X^k, Y_k)$ is assigned to the node P^k of $H(E)$ corresponding to the pair (X^k, Y_k) with $(0 \leqq k \leqq |V|-1)$. P^0 is assigned to the system function $\varphi(\underline{z})$. If in a pair (X^k, Y_k) the second set Y_k is a cut of $\underline{S}$ then $F(\underline{z}, X^k, Y_k)=0$ is valid, i.e. you cannot generate a spanning tree from $T(X^k)$ if you exclude the elements (arcs) of Y_k. Naturally, according to Lemma 3, the corresponding nodes are assigned to functions which are equal to 0.

Theorem 3: *If $P_1^k, P_2^k, \ldots, P_l^k$ are the nodes with the level k of $H(E)$ and $F(\underline{z}, X_i^k, Y_{k,i}) \neq 0$, then these functions $(i=1, \ldots, l)$ are pairwise orthogonal and for each level $k=1, \ldots, |V|-1$*

$$\varphi(\underline{z})=F(\underline{z}, X_1^k, Y_{1,k})+\ldots+F(\underline{z}, X_l^k, Y_{l,k}) \quad (17)$$

holds.

Proof: If $C(\varnothing, \varnothing)=\{v_1^0, \ldots, v_r^0\}$ then according to Theorem 2 we obtain $\varphi(\underline{z})= =F(\underline{z}, \varnothing, \varnothing)=F(\underline{z}, \{v_1^0\}, \varnothing)+F(\underline{z}, \{v_2^0\}, \{v_1^0\})+\ldots+F(\underline{z}, \{v_m^0\}, \{v_1^0, \ldots, v_{m_0-1}^0\})$ whereby m_0 is the smallest index for which the second set $\{v_1^0, \ldots, v_{m_0}^0\}$ is a cut. If $m_0 < r_0$ then $r_0 - m_0$ nodes which a 0 is assigned to occur. Thus the theorem is valid for $k=1$.

If (17) is correct for k it is also for $k+1$, provided that $k \leqq |V|-2$: Let $F(\underline{z}, X_i^k, Y_{i,k})$ and $F(\underline{z}, X_j^k, Y_{j,k})$ be two arbitrary orthogonal Boolean products of (17) and X_i^k contain an element $v \in N$, with $v \in Y_{j,k}$. If Theorem 2 is applied to $F(\underline{z}, X_i^k, Y_{i,k})$, then we obtain orthogonal Boolean functions for $k+1$ which include v in their first set. Analogous to this with an orthogonal decomposition of $F(\underline{z}, X_j^k, Y_{j,k})$ all the second sets will contain v. Consequently, all Boolean functions are also orthogonal for $k+1$. If Theorem 2 is applied to every Boolean function in (17) the proposition of the theorem for $k+1$ follows from this. Starting from $k=1$ these conclusions can be repeated until $k=|V|-1$ is obtained. In this case all sets of candidates are empty since the sets $X^{|V|-1}$ are sets of arcs of the spanning trees. By this the theorem is proved. ■

For $k=|V|-1$ this theorem provides the propositions put forward at the beginning, since the sets of arcs of the spanning trees are minimal paths of $\underset{\sim}{S}$ for which $\varphi(\underset{\sim}{z})=1$ holds:

$$\varphi(\underset{\sim}{z})=\bigwedge_{i\in X_1} z_i\cdot\bigwedge_{j\in Y_1}\bar{z}_j+\ldots+\bigwedge_{i\in X_l} z_i\cdot\bigwedge_{j\in Y_l}\bar{z}_j$$

with X_i being the sets of arcs of the spanning trees with the root s ($i=1, \ldots, l$).

Example 1: The t-graph depicted in Figure 1 with the source $s=1$ is interpreted as a binary coherent system $\underset{\sim}{S}$ with the elements $e_1, e_2, \ldots, e_7$. The system function

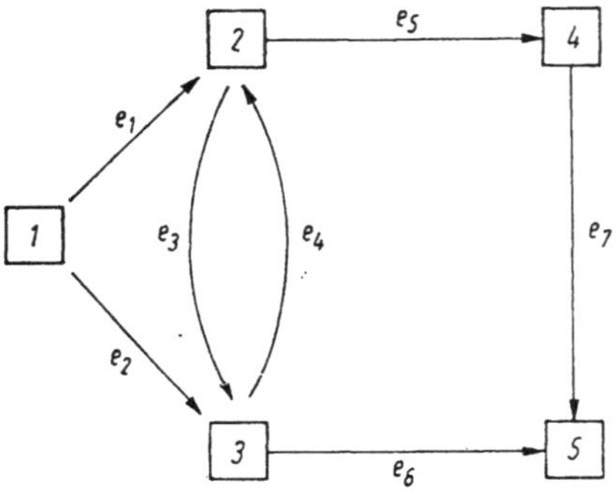

Fig. 1

of $\underset{\sim}{S}$ has to be determined as the sum of pairwise orthogonal Boolean products. The functioning of the algorithm (A1) can be seen in the tree $H(E)$ in Figure 2. We get

$$\varphi(\underset{\sim}{z})=z_1z_2z_5z_6+z_1z_2z_5z_7\bar{z}_6+z_1z_3z_5z_6\bar{z}_2+z_1z_3z_5z_7\bar{z}_2\bar{z}_6+z_2z_4z_5z_6\bar{z}_1+$$
$$+z_2z_4z_5z_7\bar{z}_1\bar{z}_6\,.$$

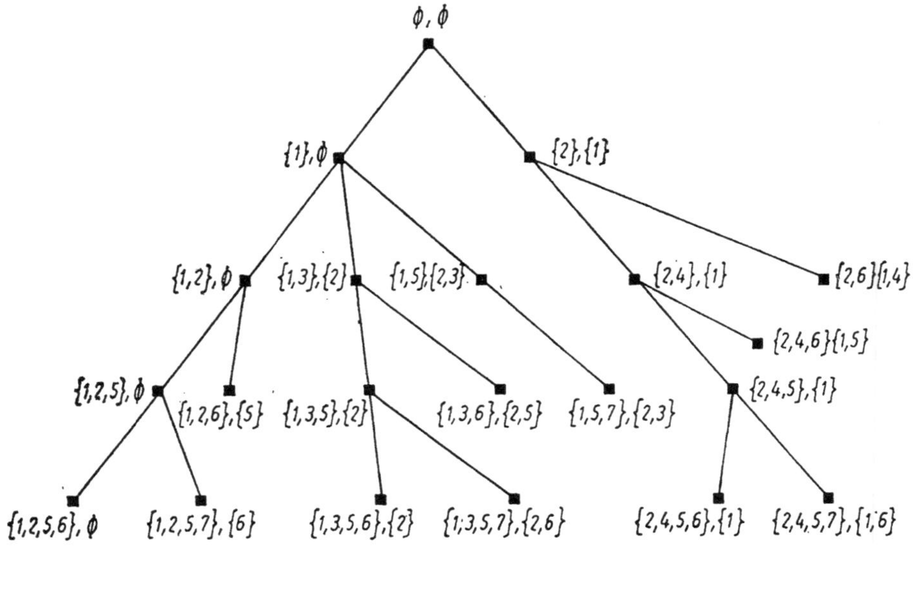

Fig. 2

The sets of arcs of the spanning trees are:

$$\{e_1, e_2, e_5, e_6\}, \{e_1, e_2, e_5, e_7\}, \{e_1, e_3, e_5, e_6\},$$
$$\{e_1, e_3, e_5, e_7\}, \{e_2, e_4, e_5, e_6\}, \{e_2, e_4, e_5, e_7\}.$$

Note: The algorithm (A1) can be applied to undirected connected graphs. The edges, being incident to s are considered to be arcs with the initial node s. All of the remaining edges are replaced by a pair of antiparallel directed arcs. For the formed digraph all spanning trees and all Boolean orthogonal products, respectively, are generated. With the output the pairs of arcs are again summarized in undirected edges and the original state variables are assigned to them.

5.3. Minimal Cuts in Path Graphs

Let a p-graph $G=(V, N)$ be given with the source s and the terminal t. A *path graph* is a graph with the following properties: Any arc in G belongs to at least one of the simple paths leading from s to t. s is the node basis, and t is the contra node basis of G. Then it is possible to consider G as a binary coherent system $\underset{\sim}{S}$ whose elements e_i $(i=1, 2, \ldots, n)$ correspond to the arcs of G.

Let the nodes be absolutely reliable. $\underset{\sim}{S}$ is referred to as operating if there is at least one simple path leading from s to t the arcs of which are operating. The arcs of a simple path of G leading from s to t are the minimal paths of $\underset{\sim}{S}$. All minimal cuts of $\underset{\sim}{S}$ are searched.

In this case a *cut* is a subset of the set of arcs N which contains at least one arc of each simple path leading from s to t. A *minimal cut* has no strict subset which is cut itself. This definition does not answer the corresponding term of cut in the sense of the graph theory. The minimal cuts in the sense of the graph theory are called *cocycles* [4] in the following.

The given algorithm (A1)′ is suited as well for the generation of all minimal cuts of $\underset{\sim}{S}$ as for the determination of all cocycles of G, being introduced later on. There are efficient algorithms [11] for undirected nonseparable graphs without s-t-path isolated nodes aimed at generating all minimal cuts based on the following theorem:

Theorem 4: *A minimal cut of G corresponds to a partition of the set of nodes V into two subsets X, $V\backslash X$ with $s\in V$ and $t\in V\backslash X$, each generating two connected subgraphs $G(X)=(X, N_X)$ and $G(V\backslash X)=(V\backslash X, N_{V-X})$ in G.*

The needed minimal cuts are determined by generating all pairs of sets $(X, V\backslash X)$ satisfying the assumptions of this theorem.

This method cannot immediately be transferred to path graphs.

Sets of nodes: Any minimal cut of $\underset{\sim}{S}$ decomposes V into three well-determined subsets:

N_u is the subset of V with the properties:

$s\in N_u$, a node $p\in V$, $p\neq s$ is contained in N_u, if there is a path from s to p which does not contain an arc of the cut S.

N_t is the subset of V with the properties:

$t \in N_t$, a node $p \neq t$ is contained in N_t if there is a path from p to t which does not contain an arc of the cut S.

$N_b = V \backslash N_u \backslash N_t$ is either empty or contains nodes of V, which can neither be reached from s nor from which t can be reached without including any arc of S.

Sets of arcs: For $X \subset V$ the notation $R^+(X)$ describes the set of all arcs the initial nodes of which are included in X and the terminal nodes of which are included in $V \backslash X$. Their terminal nodes are *neighbours* of X. $R^-(X)$ is the set of all arcs the terminal nodes of which are contained in X but whose initial nodes are not contained in X. The set of arcs $R(X) = R^+(X) \cup R^-(X)$ is the *cocycle* of X.

System of sets E: The system consists of subsets X from V with the properties: $\{s\} \in E$. If $X \in E$ and if $q \neq t$ is a neighbour of X, then $X + \{q\} \in E$. E is defined in such a way that it satisfies Conditions 1 and 2. Thus, (A1) generates all elements of E.

If $X \in E$ there is a well-determined set $V(X)$ with the properties: $t \in V(X)$. A node p being different from t is element of $V(X)$ if there is a path leading from p to t which does not contain a node of X.

Theorem 5: *$S = R^+(X) \cap R^-(V(X))$ is a minimal cut of $\underline{S}$ for any $X \in E$.*

Proof: If $p \in V \backslash X \backslash V(X)$ then a neighbour of p is either contained in this set again or is contained in X. Any path leading from s to t has an arc (pq) with $q \in V(X)$ and $p \in X$. So any path leading from s to t has an arc which is contained in $R^+(X) \cap R^-(V(X))$. Consequently, S is a cut.

S is even a minimal cut since for every arc $(pq) \in S$ there is a path leading from s to t via (pq). Therefore there is no strict subset of S, which is a cut. ∎

Conclusions:

$$S = R^-(V(X)), \tag{19}$$

$$S = R^+(N_u). \tag{20}$$

There is a minimal cut S of $\underline{S}$ for every pair $(X, V(X))$. $V(X)$ corresponds to N_t. In general, several pairs occur corresponding to the same minimal cut.

It would already be possible to give an algorithm, which generates all elements of E and in every case if X corresponds to N_u the algorithm puts out the arcs of the minimal cut according to (20). This algorithm has the disadvantage that also all X for which $X \neq N_u$ holds are generated. A more efficient algorithm (A1)′ may be given by modifying (A1). It will be better to use the following restricted definition of the term *candidate* in algorithm (A1). For an element $X \in E$ the terminal nodes of the minimal cut $S = R^-(V(X))$ being different from t are candidates.

We have to show that all minimal cuts are generated despite of this modification. By this we still consider the system of sets E despite the fact that not all its elements are generated.

Definition 5: If a minimal cut S with the set N_u is given all nodes of N_u, not being initial nodes of an arc of S, are called *nodes isolated of the cut S.*

Lemma 4: *If $X_1, X_2 \in E$ and $X_1 \neq X_2$ holds, then the pairs of sets $(X_1, V(X_1))$ and $(X_2, V(X_2))$ exactly correspond to the same minimal cut S iff X_1 and X_2 only differ in nodes of N_u which are isolated from S.*

Proof: If the pairs of sets are different and correspond to the same minimal cut then $S = R^-(V(X_1)) = R^-(V(X_2))$ holds. Thus, all arcs of S are well-determined. Consequently, X_1 and X_2 contain the initial nodes of S. If X_1 and X_2 only differ in nodes isolated from the minimal cut then obviously the same minimal cut belongs to them. ■

Lemma 5: *Each minimal cut S can be generated by the candidates defined above.*

Proof: Let the well-determined set of nodes N_u belong to S. If p' is the initial node of an arc of S, then there is a path leading from s to p' the nodes of which belong to N_u. If you consider the first arc (sp) of this path, then p, as a terminal node of the minimal cut belonging to s, is a candidate of s. If X is the set of the first k nodes of the path and if p' is not yet reached, then the minimal cut S_X belonging to X contains an arc (pq) of the path. So q is a candidate of X and $X + \{q\}$ is a continuation of the path. This construction is repeated until $q = p'$ holds. By this for any initial node p' of an arc of S a path the nodes of which are contained in N_u may be found. ■

The problem of the multiple generation of a minimal cut is dealt with by the following Lemma.

Lemma 6: *Let a minimal cut $S = R^-(V(X^k))$ be given. The condition $Y_k \cap N_u \neq \emptyset$ is necessary and sufficient for the case that the minimal cut S belonging to X^k has already been generated by* (A1).

Proof: Assume that there is a node $q \in Y_k$ with $q \in N_u \backslash X^k$ ($Y_k \cap X^k = \emptyset$ is always satisfied). From $q \in Y_k$ follows that there is an element $X^i \in E$ with $X^i \subset X^k$ which contains q in its reduced set of candidates and from which X^k is generated excluding q.

There is an element $X' = X^i + \{q\}$ in E. If any initial node p' of S is not contained in X' the construction of the previous Lemma is applied in order to generate an element X'' of E corresponding to the same cut S. Since X'' was generated by using node q, but X^k is generated excluding q, X'' must be generated before generating X^k.

Now let X^j and X^k belong to the same minimal cut. Let X^j be generated before X^k. Then there will be an $X^i = X^j \cap X^k$, the set of candidates of which contains q and from which X^j is generated by using q and from which X^k is generated with $q \in Y_k$. Since X^j and X^k correspond to the same minimal cut S, according to Lemma 4 they only differ from each other in nodes of N_u which are isolated from S. That's why $q \in N_u \backslash X^k$ holds. ■

Algorithm (A1)' is formed from (A1) by the following modifications:

- The already mentioned restriction of the definition of candidate is used.
- By means of the Boolean function *test* the repeated output of the same minimal cuts is avoided.

Algorithm (A1)′

```
k:=0; X^0:={s}; Y_0:=∅; kand (X^0, Y_0); output (S);
while k>=0 do
begin if  C(X^k, Y_k)≠∅ then
      begin
          choose (X^k, Y_k, v_k);
          X^{k+1}:=X^k+{v_k}; Y_{k+1}:=Y_k; k:=k+1;
          if test (Y_k, N_u) then
             begin
                 kand (X^k, Y_k);
                 output (S)
             end
          else C(X^k, Y_k):=∅;
      end
          else
      begin
          k:=k-1;
          if k>=0   then   Y_k:=Y_k+{v_k};
      end;
end; {while}
```

Notes: – test (Y, N_u) is a Boolean function procedure which is exactly true iff $N_u \cap Y = \emptyset$ holds. N_u is the set of nodes which, according to (20), is assigned to the cut S belonging to X^k. There are various simple possibilities to determine test.

– kand (X, Y) generates the set of candidates $C(X, Y)$.

– choose (X, Y, v) is the procedure already defined in (A1).

Elementary cocycles: Let a digraph $G=(V, N)$ be given. Assume $G_1=(V_1, N_1)$ and $G_2=(V_2, N_2)$ are two connected subgraphs of G with

$$V_1 \cap V_2 = \emptyset \quad \text{and} \quad N_1 \cap N_2 = \emptyset .$$

If the set N_{12} of all arcs of G which connect the nodes of V_1 with the nodes of V_2 is not empty, so that the graph $(V_1 \cup V_2, N_1 \cup N_2 \cup N_{12})$ is connected, then N_{12} is referred to as an *elementary cocycle* of G. An elementary cocycle contains two subsets, i.e. the set of all arcs (pq) with $p \in V_1$ and $q \in V_2$ and the set of all arcs (pq) with $p \in V_2$ and $q \in V_1$. One of these subsets may be empty.

Let $G=(V, N)$ a path graph with the source s and the terminal t there is a relationship between the minimal cuts of $\underline{S}$ and the elementary cocycles of G. Considering the pairs (G_1, G_2) with $s \in V_1, t \in V_2, N_{12} \neq \emptyset$, V_1 and V_2 correspond to the sets X and $V(X)$. Primarily, for cocycles

$$N_{12} = R(V(X)) , \tag{21}$$

$$N_{12} = R(N_u) \tag{22}$$

hold.

Therefore an elementary cocycle which additionally contains the arcs (pq) with $p \in V(X)$ and $q \in X$ corresponds to a well-determined minimal cut S. It is not possible that an elementary cocycle does not correspind to a minimal cut of $\underline{S}$ because G is a path graph. Therefore algorithm (A1)′ is suited for the generation of all elementary cocycles. In addition to the candidates according to (22) or (21) the procedure kand (X, Y) must determine the terminal nodes of the arcs belonging to the elementary cocycle. The procedure output must put out all arcs belonging to the cocycle.

Example 2: The graph G illustrated in Figure 3 has the source $s=1$ and the terminal $t=7$. $X=\{1, 4, 3\}$ is an element of $\dot{E}$ with $N_u=\{1, 2, 3, 4\}$, $V(X)=\{5, 7\}$, $N_b=\{6\}$. The minimal cut $\{e_2, e_9\}$ belongs to the pair $(X, V(X))$.

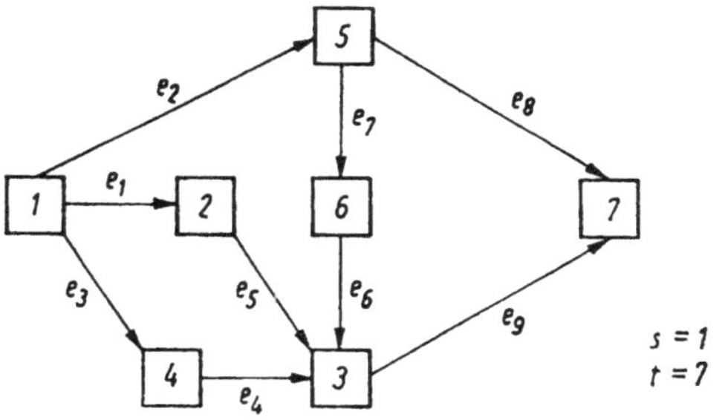

Fig. 3

Example 3: The graph G of Example 2 is interpreted as a binary coherent system $\underline{S}$. The elements of $\underline{S}$ are $e_1, e_2, \ldots, e_9$. All minimal cuts of the system are searched. The elements X and the minimal cuts belonging to them are given so that the method of working of (A1)′ may be followed up step by step.

elements of X	elements of S
1	e_1, e_2, e_3
1, 2	e_2, e_3, e_5
1, 2, 3	e_2, e_9
1, 2, 3, 5	e_8, e_9
1, 2, 4	e_2, e_4, e_5
1, 2, 4, 5	e_4, e_5, e_7, e_8
1, 2, 4, 5, 6	e_4, e_5, e_6, e_8
1, 2, 5	e_3, e_5, e_7, e_8
1, 2, 5, 6	e_3, e_5, e_6, e_8
1, 4	e_1, e_2, e_4
1, 4, 3	no new cut
1, 4, 5	e_1, e_4, e_7, e_8
1, 4, 5, 6	e_1, e_4, e_6, e_8
1, 5	e_1, e_3, e_7, e_8
1, 5, 6	e_1, e_3, e_6, e_8
1, 5, 6, 3	no new cut

References

[1] Arunkumar, S.; S. H. Lee: Enumeration of all minimal cut-sets for a node pair in a graph. IEEE Trans. Reliability, **28** (1979) 1, 51–55.

[2] Barlow, R. E.; F. Proschan: Mathematical Theory of Reliability. John Wiley and Sons, New York 1965.

[3] Beichelt, F.; P. Franken: Zuverlässigkeit und Instandhaltung. VEB Verlag Technik, Berlin 1983.

[4] Berge, C.; A. Ghoula-Houri: Programme, Spiele, Transportnetze. BSB. G. Teubner Verlagsgesellschaft, Leipzig 1969.

[5] Esary, I. D.; F. Proshan: The (Reliability of Coherent Systems. R. H. Wilcox and C. W. Mann (eds.) Redundancy Techniques for Computing Systems. Washington, D. C.: Spartan Books, 47–61.

[6] Hopcroft, J.; R. Tarjan: Algorithm 447 Efficient algorithms for graph manipulation (H). Communications of the ACM, **16** (1973) 6, 372–378.

[7] Nichols-Hagstrom, J.: Combinatoric tools for computing network reliability. Ph. D. Thesis, Dept. of IEOR, University of California, Berkeley, 1980.

[8] Read, R. C.; R. E. Tarjan: Bounds on Backtrack Algorithms for Listing Cycles, Path and Spanning Trees. Networks, **5** (1975) 237–252.

[9] Satyanarayana, A.; J. N. Hagstrom: A new Formula and an Algorithm for the Reliability Analysis of Multi-Terminal Networks. University of California, Barkeley, June 1980, ORC 80–11.

[10] Torrey, J.: A Pruned Tree Approach to Reliability Computation. IEEE Transactions on Reliability, **32** (1983) 2, 170–174.

[11] Tsukiyama, S.; H. Ariyoshi; I. Shirakawa; H. Ozaki: An Algorithm to Enumerate All Cutsets of a Graph. JACM, **27** (1980) 4, 619–632.

Received July 1987

F. Zettwitz
Ingenieurhochschule Mittweida
Sektion Informationselektronik
Platz der DSF 17
Mittweida
DDR - 9250

optimization 19 (1988) 6, 889–894

Improved Lower and Upper Bounds for the Number of Feasible Solutions to a Knapsack Problem

M. Hujter

Hungarian Academy of Sciences, Computer and Automation Institute
Budapest

Summary: Some known results about lower and upper bounds for the number of distinct solutions to a discrete knapsack problem are given. In this paper some sharper bounds are proved by using geometrical methods. The proof of the upper bound is based on the famous (and deep) Brunn-Minkowski inequality, and it is a new and interesting application method of geometrical arguments in discrete programming.

AMS 1980 Subject Classification:

Primary: 90 C 10

Key words: Discrete programming, knapsack problem, geometrical methods, Brunn-Minkowski inequality.

1. Introduction

Let $w_1, w_2, \ldots, w_n$ and W be fixed positive integers and a positive real number, respectively, and let N denote the number of distinct solutions to the following (discrete) *knapsack problem*:

$$\sum_{j=1}^{n} w_j x_j \leqq W; \quad x_j \geqq 0 \quad \text{and integer}, \quad j = 1, 2, \ldots, n .$$

It would be nice to have a simple formula for N in terms of $w_1, w_2, \ldots, w_n$ and W. If f_m denotes the coefficient of x^m in the power series development of the function

$$\Phi(x) = \prod_{j=1}^{n} \frac{1}{1 - x^{w_j}}$$

then

$$N = \sum_{m=0}^{W} f_m$$

which is the same as the coefficient of $x^{[W]}$ in the power series development of the function

$$\Psi(x) = \frac{\Phi(x)}{1 - x}.$$

Using these facts, O. Achou [1] could derive the following formula for N if the numbers $w_1, w_2, \ldots, w_n$ are pairwise relatively prime:

$$N=\sum_{p=0}^{n}\binom{[W]+p}{p}N_p+\sum_{j=1}^{n}\sum_{r=1}^{w_j-1}D_{rj}/\varepsilon_j^{r(W+1)}$$

where

$$\varepsilon_j=\exp\,(2\pi i/w_j)\quad(j=1,\ldots,n)$$

$$D_{rj}=\frac{\varepsilon_j^r}{w_j\,(1-\varepsilon_j^r)}\Big/\prod_{\substack{k=1\\k\neq j}}^{n}(1-\varepsilon_j^{rw_k})\quad(j=1,\ldots,n;\ r=1,2,\ldots,w_j-1)$$

and

$$N_p=-\sum_{j=1}^{n}\sum_{r=1}^{w_j-1}D_{rj}\,(1-\varepsilon_j^r)^p\quad(p=0,1,\ldots,n)\,.$$

Because of the precedence of complex roots of unity, this formula is awkward to calculate numerically. J. Piehler [5] transformed this formula so that the calculation can process using only rational numbers. However, even in this form we need more than S additions for the calculation of N. Since no fast method to compute the exact value of N has been found, lower and upper bounds have great significance.

A. G. Beged-Dov [2] has proved that

$$W^n/P\leqq N\leqq(W+S)^n/P$$

where

$$S=\sum_{j=1}^{n}w_j\quad\text{and}\quad P=n!\prod_{j=1}^{n}w_j\,.$$

In this paper we prove the following sharper bounds:

$$(V+d)^n/P\leqq N\leqq\left(2\left(V+\frac{1}{2}S\right)^n-V^n\right)\Big/P$$

where d denotes the greatest common divisor of $w_1, \ldots, w_n$, and $V=d\cdot[W/d]$ (i.e. the largest integer which is not larger than W and a multiple of d). Since $V\leqq W$, $V+d>W$ and (from the convexity of the power function)

$$2\left(V+\frac{1}{2}S\right)^n\leqq V^n+(V+S)^n\,,$$

our bounds are in fact improvements of the bounds of Beged-Dov [2].

Our lower bound is proved by the methods of Padberg [4] in Section 2. The upper bound requires some kind of geometric arguments. We show this method in Section 3. Finally, in Section 4 we provide the reader a table of the old and the new lower and upper bounds for some simple configurations.

2. Improved Lower Bound

Theorem 1: $(V+d)^n/P \leqq N$.

Proof: For any $W>0$, let $A(W)$ denote the closed set in $\mathbb{R}^n$ determined by $n+1$ inequalities as follows:

$$\underline{x} \in A(W) \Leftrightarrow \begin{cases} x_j \geqq 0 \quad (j=1, \ldots, n) \\ \sum_{j=1}^{n} w_j x_j \leqq W . \end{cases}$$

Actually, $A(W)$ is a simplex with vertices $(0, 0, \ldots, 0)$, $(W/w_1, 0, \ldots, 0)$, $\ldots$, $(0, 0, \ldots, 0, W/w_n)$. It is well-known that the n-dimensional volume of the simplex with vertices $(0, 0, \ldots, 0)$, $(a_1, 0, \ldots, 0)$, $(0, a_2, 0, \ldots, 0)$, $\ldots$, $(0, \ldots, 0, a_n)$ is equal to

$$\left(\prod_{j=1}^{n} a_j\right)\Big/ n! .$$

Thus the n-dimensional volume of $A(W)$ is equal to W^n/P.

Now for any $1>\varepsilon>0$ consider the simplex $A\ (V+d-\varepsilon)$. Observe that for any integer vector $\underline{x}=(x_1, \ldots, x_n)$,

$$\underline{x} \in A(W) \Leftrightarrow \underline{x} \in A\ (V+d-\varepsilon) .$$

Now for all such integer vectors $\underline{x}=(x_1, \ldots, x_n)$, consider the sets

$$B(\underline{x})=B(x_1, \ldots, x_n)=A\ (V+d-\varepsilon) \cap \mathop{\times}_{j=1}^{n} [x_j, x_j+1) .$$

These sets determine a partition of $A\ (V+d-\varepsilon)$. Obviously, the n-dimensional volume of each $B(\underline{x})$ is at most 1, and the number of the sets is N. Therefore

$$N \geqq \frac{\text{the volume of } A\ (V+d-\varepsilon)}{\text{the maximum volume of } B\ (\underline{x})} \geqq (V+d-\varepsilon)^n/P .$$

Now if $\varepsilon \to 0$, we obtain Theorem 1. ■

3. Improved Upper Bound

Theorem 2: $N \leqq \left(2\left(V+\frac{1}{2}S\right)^n - V^n\right)\Big/ P$.

Proof: First of all observe that each of the unit cubes

$$C(\underline{x})=C(x_1, \ldots, x_n)=\mathop{\times}_{j=1}^{n} [x_j, x_j+1)$$

is contained in $A\ (V+S)$ for all integer vectors

$$\underline{x}=(x_1, \ldots, x_n) \in A(W) .$$

Obviously, these unit cubes are pairwise disjoint. Let $U\ (V+S)$ denote their union.

The volume of $U\,(V+S)$ is exactly N. Observe that the center of each $C(\underline{x})$ is in $A\left(V+\frac{1}{2}S\right)$.

Now we will apply the famous BRUNN-MINKOWSKI inequality. Denote by $X+Y$ the algebraic sum of two sets X, $Y\in\mathbb{R}^n$. The BRUNN-MINKOWSKI inequality asserts that

$$\mu_{n*}\,(\lambda X+(1-\lambda)\;Y)\geqq(\lambda\mu_n(X)^{1/n}+(1-\lambda)\;\mu_n(Y)^{1/n})^n$$

where $0\leqq\lambda\leqq1$, μ_n and μ_{n*} mean LEBESQUE-measure and inner LEBESQUE-measure, respectively, and X, Y are LEBESQUE-measurable sets (see, e.g., [3]). If X and Y are convex, compact sets, then so is $\lambda X+(1-\lambda)\;Y$, and thus μ_n and μ_{n*} simply mean the n-dimensional volumes of the above sets.

In our case let

$$Y=C(\underline{x})\setminus A\left(V+\frac{1}{2}S\right)$$

if this set is nonempty, and let

$$X=\{(2x_1+1-y_1,\ldots,2x_n+1-y_n)\mid(y_1,\ldots,y_n)\in Y\}\,.$$

Hence, $X\subseteqq C(\underline{x})$ and the volume of X is the same as the volume of Y. Let furthermore

$$\lambda=\frac{\Omega-V+\frac{1}{2}S}{V-\Omega+\frac{1}{2}S}$$

where

$$\Omega=\sum_{j=1}^{n}w_jx_j\,.$$

$\left(\text{Since } Y\neq\emptyset\text{, we have } V-\frac{1}{2}S\leqq\Omega\leqq V\text{, thus } 0\leqq\lambda\leqq1.\right)$ Observe that if

$$Z=\lambda X+(1-\lambda)\;Y\,,$$

then

$$Z\subseteqq C(\underline{x})\cap\left(A\left(V+\frac{1}{2}S\right)-A(V)\right)$$

holds. (See Figure 1.)

Applying the BRUNN-MINKOWSKI inequality, we obtain that the volume of Z is at least the volume of Y. Therefore the volume of Y cannot be larger than the volume of

$$C(\underline{x})\cap\left(A\left(V+\frac{1}{2}S\right)-A(V)\right).$$

This fact is true for all cases when $Y=\emptyset$. This implies that the volume of

$$U\,(V+S)\cap\left(A\left(V+\frac{1}{2}S\right)-A(V)\right)$$

is greater than or equal to the volume of

$$U\,(V+S)\cap\left(A\,(V+S)-A\left(V+\frac{1}{2}S\right)\right).$$

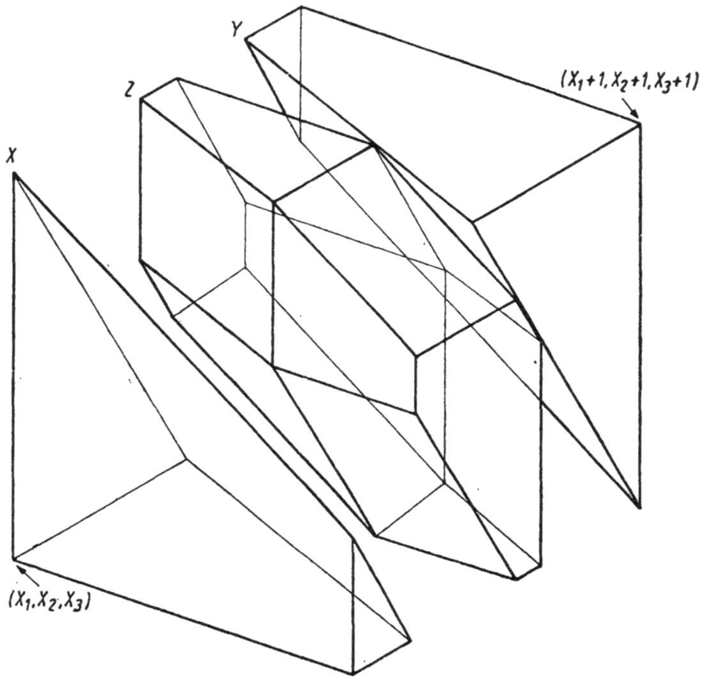

Fig. 1

Hence, the volume of $U\,(V+S)$, which is N, is at most the volume of $A(V)$ plus at most twice the volume of $A\left(V+\frac{1}{2}S\right)-A(V)$. Thus we obtain that

$$N\leqq V^n/P+2\left(\left(V+\frac{1}{2}S\right)^n/P-V^n/P\right).$$

This inequality is equivalent to Theorem 2. ■

4. Examples

In this section we provide some examples (see Table 1). The 'old' bounds are the results of BEGED-DOV [2], the 'new' bounds are our improvements. At the lower (upper) bounds we took the upper (lower) integer parts of the bounds.

Table 1

W	w_1	w_2	w_3	old lower bound	new lower bound	exact value of N	new upper bound	old upper bound
1	2	3	5	1	1	1	2	7
2	2	3	5	1	1	2	3	9
5	2	3	5	1	2	6	10	18
10	2	3	5	6	8	20	31	44
20	2	3	5	45	52	91	129	150
50	2	3	5	695	737	947	1154	1200
100	2	3	5	5556	5724	6518	7306	7394
200	2	3	5	44445	45115	48202	51279	51450
500	2	3	5	694445	698620	717587	736529	736950
1000	2	3	5	5555556	5572239	5647673	5723056	5723894

Acknowledgement

I am grateful to Imre Bárány, Béla Vizvári and Lajos Rónyai for fruitful conversations.

References

[1] Achou, Ö.: The number of feasible solutions to a knapsack problem. SIAM J. Appl. Math., **27** (1974) 606–610.
[2] Beged-Dov, A. G.: Lower and upper bounds for the number of lattice points in a simplex. SIAM J. Appl. Math., **22** (1972) 106–108.
[3] Hadwiger, H.: Vorlesungen über Inhalt, Oberfläche und Isoperimetrie. Springer, Berlin 1957.
[4] Padberg, M. W.: A remark on an inequality for the number of lattice points in a simplex. SIAM J. Appl. Math., **20** (1971) 638–641.
[5] Piehler, J.: Bemerkung zu einer Arbeit von Achou über die Anzahl zulässiger Lösungen eines Knapsack-problems. Math. Operationsforsch. Statist., Ser. Optimization, **9** (1978) 167–170.

Received October 1987

M. Hujter
Hungarian Academy of Sciences
Computer and Automation Institute
Kende u. 13–17
Budapest
Hungary H - 1111

optimization 19 (1988) 6, 895–904

Über ein Vektoroptimierungsproblem für endliche stochastische zellulare Systeme

J. STORM

Pädagogische Hochschule Dresden, Sektion Mathematik

Summary: In this note we consider a vector-valued decision problem for finite stochastic cellular systems, based on the PARETO-minimum of vectors, components of which are average costs corresponding to the cells of the system. We give a policy-improvement algorithm and show, under which conditions a policy is optimal.

Zusammenfassung: In der Arbeit wird ein vektorwertiges Entscheidungsproblem für endliche stochastische zellulare Systeme betrachtet, das auf dem PARETO-Minimum der Vektoren aus den Durchschnittskosten bez. jeder Zelle des Systems basiert. Es wird ein Verfahren zur Strategieverbesserung angegeben und gezeigt, unter welchen Bedingungen eine Strategie optimal ist.

AMS 1980 Subject Classification:

Primary: 90 B 50; Secondary: 93 E 20.

Key words: Multi-objective stochastic decision model, control of stochastic processes, stochastic cellular system.

0. Einleitung

Zahlreiche real existierende Systeme, wie z. B. biologische Zellkomplexe, homogene mikroelektronische Strukturen oder digital aufgebaute Systeme der Bildverarbeitung zeichnen sich durch eine große Anzahl gleichartiger Elemente (Zellen) aus, die in Form eines regelmäßigen, n-dimensionalen Gitters ($n = 1$, 2, 3) angeordnet sind, wobei jede Zelle mit einer gewissen Teilmenge der Menge aller Zellen in Wechselwirkung steht.

Stochastische zellulare Systeme (SZS) (vgl. z. B. [1]) stellen ein abstraktes Modell dieser realen Systeme dar. Gegenstand dieser Arbeit ist die Untersuchung eines speziellen Steuerungsproblems für endliche (eindimensionale) SZS. Jeder Zelle des Systems wird eine reellwertige Kostenfunktion zugeordnet. Ein solches SZS kann als ein stochastisches dynamisches Entscheidungsmodell interpretiert werden (vgl. hierzu [2]). Als Optimalitätsziel betrachten wir die Minimierung des Vektors der Durchschnittskosten (pro Takt) jeder Zelle im Sinne des PARETO-Minimums. In Anlehnung an die Arbeiten von THOMAS [3] und FURUKAWA [4] werden in Abschnitt 2 Eigenschaften dieses multikriteriellen Entscheidungsmodells untersucht. Ein Verfahren zur Strategieverbesserung wird in Abschnitt 3 vorgestellt. Abschließend geben wir in Abschnitt 4 eine Charakterisierung für optimale Strategien an und stellen einen Algorithmus zur Bestimmung der minimalen Kosten vor.

1. Modellbeschreibung

Wir beschränken uns im folgenden auf die Untersuchung eindimensionaler SZS mit endlichem Systemraum. Endliche n-dimensionale SZS können in analoger Weise behandelt werden. In jedem Punkt r des (eindimensionalen) *Systemraumes* $R=\{1, 2, \ldots, m\}$ befinde sich eine *stochastische Zelle* von demselben Typ (X, Z, Y, P, φ). Dabei sind X, Z, Y endliche Mengen mit $Y \subset \mathbb{R}$ und bezeichnen entsprechend *Eingabe-*, *Zustands-* und *Ausgabealphabet* der Zelle. Jede Zelle r ist mit den Zellen ihrer *Nachbarschaft* $N(r)$ gekoppelt. Die Art der Kopplung sei für alle Zellen dieselbe und werde durch den *Nachbarschaftsindex* $N=\{n_0, n_1, \ldots, n_s\}$ beschrieben, wobei $n_0=0$, $n_i \in \mathbb{Z}\backslash\{0\}$, $i=1, \ldots, s$. Die Nachbarschaft einer Zelle r ist dann durch $N(r)=\{r, r+n_1, \ldots, r+n_s\}$ bestimmt. Es bezeichne $\dot{Z}:=\{\dot{z} \mid R \to Z\}$, $\dot{X}:=\{\dot{x} \mid R \to X\}$ entsprechend die Mengen aller *Zustands-* bzw. *Eingabekonfigurationen* des Systems. $\dot{z}[r]$, $\dot{x}[r]$ und $\dot{z}_{N(r)}$ geben jeweils den Zustand und das Eingabesignal der Zelle r bzw. die Gesamtheit aller Zustände der Nachbarschaft $N(r)$ an, wenn sich das System in dem (globalen) Zustand $\dot{z}$ befindet und das (globale) Eingabesignal $\dot{x}$ erhält.

Das *lokale Übergangsverhalten* einer Zelle wird durch einen (lokalen) stochastischen Kern $P(.|.)$ von $\dot{Z}_N \times X$ nach Z beschrieben, wobei $\dot{Z}_N:=\{\dot{z}_r=(z_0, z_1, \ldots, z_s) \mid z_0 \in Z, z_i \in Z \cup \{\Lambda\}, i=1, 2, \ldots, s\}$ und Λ ein „Leerzustand" ist. Seine Einführung ergibt sich aus der Gleichförmigkeit der Kopplung und der Beschränktheit des Systemraumes. Es wird vereinbart, daß $\dot{z}[r+n_i]=\Lambda$ für $r+n_i \notin R$, $1 \leq i \leq s$ gesetzt wird. $\varphi: Z \to Y$ ist eine reellwertige *Ausgabefunktion*. Somit besitzt jede Zelle einen externen Ein- und Ausgang. Alle Zellen des Systems arbeiten getaktet und synchron.

Definition 1: Das Tripel $\mathfrak{A}=\langle R, N, (X, Z, Y, P, \varphi)\rangle$ wird *stochastisches zelzulares System* genannt.

Unter Verwendung der in der Theorie der Steuerung stochastischer Prozesse üblichen Terminologie interpretieren wir $\dot{X}$ als *Steuerraum*, $\dot{Z}$ als *Zustandsraum* und $H_t:=\dot{Z}\times\dot{X}\times\dot{Z}\times \ldots \times\dot{Z}$ ($2t+1$ Faktoren) als die Menge der *Vorgeschichten* bis zum Zeitpunkt $t \in \mathbb{N}$ ($\mathbb{N}=\{0, 1, 2, \ldots\}$). Als *Strategie* $\sigma=(\nu_t)_{t\in\mathbb{N}}$ bezeichnen wir eine Folge von Wahrscheinlichkeitsverteilungen $\nu_t(. \mid h_t)$ auf $\dot{X}$, $h_t \in H_t$. Die Menge aller dieser Strategien werde mit Σ bezeichnet. Hängen die Verteilungen ν_t nur von $\dot{z}_t \in \dot{Z}$ und nicht von der gesamten Vorgeschichte h_t ab, so sprechen wir von einer Markovschen *Strategie*. Σ^M, Σ^S, Σ^D seien entsprechend die Mengen aller Markovschen, stationären bzw. determinierten Strategien.

Das *globale Übergangsverhalten* eines SZS $\mathfrak{A}$ ist durch einen (globalen) stochastischen Kern $\dot{P}(.\mid.)$ von $\dot{Z}\times\dot{X}$ nach $\dot{Z}$ bestimmt, wobei

$$\dot{P}(\dot{z}' \mid \dot{z}, \dot{x}) = \prod_{r\in R} P(\dot{z}'[r] \mid \dot{z}_{N(r)}, \dot{x}[r]) , \quad \dot{z}, \dot{z}' \in \dot{Z} , \quad \dot{x}\in\dot{X} , \tag{1}$$

ist. Als Übergangswahrscheinlichkeit gibt $\dot{P}(\dot{z}' \mid \dot{z}, \dot{x})$ die Wahrscheinlichkeit an, mit der $\mathfrak{A}$ zur Zeit $t+1$ den Zustand $\dot{z}'$ annimmt, wenn sich das System zur Zeit t im Zustand $\dot{z}$ befand und die Steuerung $\dot{x}$ erhielt.

Jede Strategie $\sigma=(\nu_t)_{t\in\mathbb{N}}$, mit der $\mathfrak{A}$ gesteuert wird, erzeugt bekanntlich (vgl. z. B. [5]) auf einem Wahrscheinlichkeitsraum $[\Omega, \mathfrak{F}, \mathsf{P}_\sigma]$ einen (*globalen*) *stochastischen Prozeß* $(\dot{\mathfrak{Z}}_t, \dot{\mathfrak{X}}_t)_{t\in\mathbb{N}}$, wobei $\Omega=\mathop{\times}\limits_0^\infty (\dot{Z}\times\dot{X})$, $\mathfrak{F}=\bigotimes\limits_0^\infty (\mathfrak{P}(\dot{Z})\otimes\mathfrak{P}(\dot{X}))$, $\dot{\mathfrak{Z}}_t(\omega)=\dot{z}_t$, $\dot{\mathfrak{X}}_t(\omega)=\dot{x}_t$ für $\omega=((\dot{z}_0, \dot{x}_0), (\dot{z}_1, \dot{x}_1), \ldots)\in\Omega$ gilt. Das Wahrscheinlichkeitsmaß P_σ ist durch

$$\begin{aligned}&\mathsf{P}_\sigma(\dot{\mathfrak{Z}}_0=\dot{z}_0, \dot{\mathfrak{X}}_0=\dot{x}_0, \ldots, \dot{\mathfrak{X}}_{t-1}=\dot{x}_{t-1}, \dot{\mathfrak{Z}}_t=\dot{z}_t)\\ &=\dot{p}_0(\dot{z}_0)\prod_{n=0}^{t-1} \nu_n(\dot{x}_n \mid h_n)\, \dot{P}(\dot{z}_{n+1} \mid \dot{z}_n, \dot{x}_n)\end{aligned}$$

definiert. Dabei sind $h_t=(\dot{z}_0, \dot{x}_0, \ldots, \dot{x}_{t-1}, \dot{z}_t)\in H_t$, $h_n=(\dot{z}_0, \dot{x}_0, \ldots, \dot{x}_{n-1}, \dot{z}_n)$, $0\leqq n\leqq t-1$, $t\in\mathbb{N}$, und $\dot{p}_0(.)$ eine Initialverteilung auf $\dot{Z}$.

Befindet sich das System $\mathfrak{A}$ im Zustand $\dot{z}$, so gibt $\varphi(\dot{z}[r])$ die der Zelle $r\in R$ zugeordneten *Kosten* an. Es sei bemerkt, daß Funktionen vom Typ $\varphi\colon Z\to Y$, durch die eine Bewertung der Zustände vorgenommen wird, in vielen praktischen Anwendungen auftreten. Die Betrachtung solcher Kostenfunktionen stellt jedoch keine wesentliche Einschränkung für die folgenden Ausführungen dar, da sich die erhaltenen Resultate leicht auf Funktionen $\psi: Z\times X\to Y$ übertragen lassen.

Für jede Zelle $r\in R$ können wir somit eine Folge von Zufallsgrößen $(\mathfrak{W}_t(r))_{t\in\mathbb{N}}$ durch $\mathfrak{W}_t(r)=\varphi(\dot{z}[r])$ mit $\dot{\mathfrak{Z}}_t=\dot{z}$ definieren. Sei

$$\bar{d}_r(\sigma, n, \dot{z}_0) := \frac{1}{n+1}\sum_{t=0}^{n} \mathsf{E}_\sigma\,(\mathfrak{W}_t(r) \mid \dot{\mathfrak{Z}}=\dot{z}_0) ,$$

wobei unter E_σ der durch die Strategie σ erzeugte Erwartungswertoperator zu verstehen ist. $\bar{d}_r(\sigma, n, \dot{z}_0)$ gibt die mittleren Durchschnittskosten für die Zelle r bis zum Zeitpunkt n an, wenn $\dot{z}_0$ der Initialzustand des Systems $\mathfrak{A}$ ist.

Es sei ferner $d_r(\sigma, \dot{z}_0) := \limsup\limits_{n\to\infty} \bar{d}_r(\sigma, n, \dot{z}_0)$. Die Größe $d_r(\sigma, \dot{z}_0)$ stellt die *lokalen mittleren Durchschnittskosten für die Zelle* r dar. Den m-dimensionalen Vektor

$$D(\sigma, \dot{z}_0) := (d_1(\sigma, \dot{z}_0), \quad d_2(\sigma, \dot{z}_0), \ldots, d_m(\sigma, \dot{z}_0))$$

der lokalen mittleren Durchschnittskosten nennen wir *mittlere D-Kosten des Systems* $\mathfrak{A}$ für den Initialzustand $\dot{z}_0$ und die Strategie σ. Man erkennt leicht, daß das SZS $\mathfrak{A}$ mit der oben eingeführten Kostenstruktur als ein multikriterielles MARKOVsches Entscheidungsmodell interpretiert werden kann.

Das in dieser Arbeit betrachtete Optimalitätsziel basiert auf der Bildung des PARETO-Minimums einer Menge $A \subset \mathbb{R}^m$.

Definition 2: Seien $x=(x_1, \ldots, x_m)$, $y=(y_1, \ldots, y_m) \in \mathbb{R}^m$ und A eine nichtleere Teilmenge von $\mathbb{R}^m$.

a) Auf $\mathbb{R}^m$ werde eine Halbordnung durch

$$x \leqq y :\Leftrightarrow x_i \leqq y_i \quad \text{für alle } i;$$
$$x < y :\Leftrightarrow x_i \leqq y_i \quad \text{für alle } i \text{ und } x_j < y_j \text{ für wenigstens ein } j$$

definiert.

b) Die Menge $\text{Min } A := \{x \in A \mid \nexists x' \in A \text{ mit } x' < x\}$ heißt PARETO-*Minimum von A*. Wenn $\hat{x} \in \text{Min } A$, so nennen wir $\hat{x}$ ein *minimales Element* der Menge A.

Definition 3: Eine Strategie $\sigma^* \in \Sigma$ heißt *D-optimal* für den Initialzustand $\dot{z}_0 \in \dot{Z}$, wenn

$$D(\sigma^*, \dot{z}_0) \in D^*(\dot{z}_0) := \text{Min}\{D(\sigma, \dot{z}_0) \mid \sigma \in \Sigma\}$$

ist.

2. Eigenschaften des Modells

Seien $\Delta(\dot{z}) := \{D(\sigma, \dot{z}) \mid \sigma \in \Sigma\}$ und $\Delta^M(\dot{z})$, $\Delta^S(\dot{z})$, $\Delta^D(\dot{z})$, $\dot{z} \in \dot{Z}$, entsprechend die Mengen aller mittleren D-Kosten des SZS $\mathfrak{A} = \langle R, N, (X, Z, Y, P, \varphi)\rangle$, wenn auf $\mathfrak{A}$ jeweils Strategien aus Σ^M, Σ^S bzw. Σ^D angewendet werden. Die zwischen diesen Mengen bestehenden Beziehungen werden in dem folgenden Satz dargestellt, der auf THOMAS [3] zurückgeht.

Satz 1: *Sei* $\mathfrak{A} = \langle R, N, (X, Z, Y, P, \varphi)\rangle$ *ein* SZS *mit der Eigenschaft, daß der globale Zustandsprozeß* $(\dot{\mathfrak{Z}}_t)_{t \in \mathbb{N}}$ *für jede Strategie eine ergodische* MARKOV-*Kette ist. Dann gilt*

$$\Delta^D(\dot{z}) = \Delta^S(\dot{z}) = \Delta^M(\dot{z}) = \Delta(\dot{z}), \quad \dot{z} \in \dot{Z}.$$

Unter dieser sehr einschränkenden Voraussetzung an ein SZS folgt aus Satz 1, daß die minimalen mittleren D-Kosten $D^*(\dot{z})$, $\dot{z} \in \dot{Z}$, in der Menge Σ^{DS} aller stationären determinierten Strategien bestimmt werden können. Im Abschnitt 4 werden wir einen Algorithmus zur Bestimmung von $D^*(\dot{z})$ in der Menge Σ^{DS} angeben. Dazu benötigen wir die nachstehend aufgeführten Aussagen und Bezeichnungen.

Sei $\mathfrak{A} = \langle R, N, (X, Z, Y, P, \varphi)\rangle$ ein SZS, auf das eine stationäre determinierte Strategie $\sigma = (f^\infty)$, $f: \dot{Z} \to \dot{X}$, angewendet wird. Es existiere ferner eine Lösung $\{G^f, V^f(\dot{z})\}$, $\dot{z} \in \dot{Z}$, des Gleichungssystems

$$G^f + V^f(\dot{z}) = \Phi(\dot{z}) + \sum_{\dot{z}' \in \dot{Z}} \dot{P}(\dot{z}' \mid \dot{z}, f(\dot{z})) \cdot V^f(\dot{z}'), \quad \dot{z} \in \dot{Z}, \tag{2}$$

wobei $G^f = (g_1^f, \ldots, g_m^f)$, $V^f(\dot{z}) = (v_1^f(\dot{z}), \ldots, v_m^f(\dot{z})) \in \mathbb{R}^m$ und $\Phi(\dot{z}) \in \mathbb{R}^m$ durch $\Phi(\dot{z}) := (\varphi(\dot{z}[1]), \ldots, \varphi(\dot{z}[m]))$ definiert ist. Analog zu dem Vorgehen von DERMAN [6] bei Entscheidungsmodellen mit reellwertiger Kostenstruktur kann man zeigen, daß G^f die mittleren D-Kosten des Systems $\mathfrak{A}$ für die Strategie $\sigma=(f^\infty)$ sind, wobei $D(\sigma, \dot{z})=G^f$ für jedes $\dot{z}\in\dot{Z}$ gilt.

Wird gefordert, daß die lokalen Übergangswahrscheinlichkeiten $P(z \mid \dot{z}_N, x)$ für alle $z\in Z$, $\dot{z}_N\in\dot{Z}_N$, $x\in X$ streng positiv sind, so läßt sich leicht zeigen, daß folgendes Lemma gilt.

Lemma 1: *Sei* $\mathfrak{A}=\langle R, N, (X, Z, Y, P, \varphi)\rangle$ *ein* SZS *mit* $P(z \mid \dot{z}_N, x)>0$, $z\in Z$, $\dot{z}_N\in\dot{Z}_N$, $x\in X$. *Wird auf* $\mathfrak{A}$ *eine Strategie* $\sigma\in\Sigma^{DS}$ *angewendet, so ist der induzierte globale Zustandsprozeß* $(\mathfrak{Z}_t)_{t\in\mathbb{N}}$ *eine ergodische* MARKOV-*Kette.*

Wir nehmen im weiteren an, daß $\mathfrak{A}$ ein SZS mit der in Lemma 1 genannten Voraussetzung ist. Für eine beliebige Strategie $\sigma\in\Sigma^{DS}$ mit $\sigma=(f^\infty)$, $f:\dot{Z}\to\dot{X}$, lassen sich dann wegen Lemma 1 die mittleren D-Kosten G^f in der Form

$$G^f=\sum_{\dot{z}\in\dot{Z}}\mu_f(\dot{z})\cdot\Phi(\dot{z}) \tag{3}$$

darstellen, wobei $\mu_f(\cdot)$ die stationäre Verteilung auf $\dot{Z}$ ist. Sei $g:\dot{Z}\to\dot{X}$ eine beliebige Entscheidungsfunktion und $\{G^f, V^f(\dot{z})\}$, $\dot{z}\in\dot{Z}$, eine Lösung von (2). Dann setzen wir

$$\gamma_f^g(\dot{z}):=\Phi(\dot{z})+\sum_{\dot{z}'\in\dot{Z}}P(\dot{z}' \mid \dot{z}, g(\dot{z}))\cdot V^f(\dot{z}'),\quad \dot{z}\in\dot{Z}. \tag{4}$$

Lemma 2: a) *Wenn* $\gamma_f^g(\dot{z})\geqq\gamma_f^f(\dot{z})$ *für alle* $\dot{z}\in\dot{Z}$ *und* $\gamma_f^g(\dot{z})>\gamma_f^f(\dot{z})$ *für wenigstens ein* $\dot{z}$ *gilt, so ist* $G^g>G^f$.

b) *Ist* $\gamma_f^g(\dot{z})=\gamma_f^f(\dot{z})$ *für alle* $\dot{z}\in\dot{Z}$, *so ist* $G^g=G^f$.

c) *Wenn* $\gamma_f^g(\dot{z})\leqq\gamma_f^f(\dot{z})$ *für alle* $\dot{z}\in\dot{Z}$ *und* $\gamma_f^g(\dot{z})<\gamma_f^f(\dot{z})$ *für wenigstens ein* $\dot{z}$ *gilt, so ist* $G^g<G^f$.

Die Aussagen von Lemma 2 ergeben sich unmittelbar aus (3) und der Ergodizität des globalen Zustandsprozesses.

3. Ein Verfahren zur Strategieverbesserung

In [4] gibt FURUKAWA ein auf dem HOWARD-Algorithmus beruhendes Strategieiterationsverfahren für einen vektorwertigen diskontierten Entscheidungsprozeß an. Wir zeigen in diesem Abschnitt, daß sich dieses Verfahren analog auf das hier untersuchte Problem der Steuerung eines SZS übertragen läßt (vgl. Satz 2). Dazu führen wir folgende Definitionen und Bezeichnungen ein, die zum Teil [4] entnommen sind.

Mit K sei der nichtnegative Kegel in $\mathbb{R}^m$ bezeichnet, d. h. $K:=\{x=(x_1, \ldots, x_m)\in\mathbb{R}^m \mid x_i\geqq 0,\ i=1, 2, \ldots, m\}$. Seien $x\in\mathbb{R}^m$ und $A\subset\mathbb{R}^m$, dann definieren wir $x-A:=\{y\in\mathbb{R}^m \mid y=x-a,\ a\in A\}$. Die Menge aller m-Vektorfunktionen, die auf $\dot{Z}$ definiert sind, fassen wir mit $M^m(\dot{Z})$ zusammen. Für beliebige $u, v\in M^m(\dot{Z})$ definieren wir $u\leqq v$ durch $u(\dot{z})\leqq v(\dot{z})$ für alle $\dot{z}\in\dot{Z}$.

Sei I eine endliche Menge und $\{u_i\}$ eine Familie von Funktionen aus $M^m(\dot{Z})$. Die Funktion $\bigcup_{i\in I} u_i$ wird durch

$$(\bigcup_{i\in I} u_i)\ (\dot{z}) := \bigcup_{i\in I} (u_i(\dot{z}))\ ,\quad \dot{z}\in\dot{Z}\ ,$$

definiert. Ferner bezeichne S^m die Menge aller nichtleeren Teilmengen von $\mathbb{R}^m$ und $L^m(\dot{Z})$ die Menge aller Abbildungen von $\dot{Z}$ in S^m. Für jedes $U\in L^m(\dot{Z})$ sei die Funktion Min U auf $\dot{Z}$ durch

$$(\text{Min } U)(\dot{z}) := \text{Min } \{U(\dot{z})\}\ ,\quad \dot{z}\in\dot{Z}\ ,$$

definiert. Wenn Min $\{U(\dot{z})\} \neq \emptyset$ für jedes $\dot{z}$, so ist Min $U\in L^m(\dot{Z})$. Für beliebige $U,\ V\in L^m(\dot{Z})$ definieren wir die Funktion $U\cap V$ auf $\dot{Z}$ durch

$$(U\cap V)(\dot{z}) := U(\dot{z}) \cap V(\dot{z})\ ,\quad \dot{z}\in\dot{Z}\ .$$

Die Beziehung $u\in U$ für $u\in M^m(\dot{Z})$, $U\in L^m(\dot{Z})$ bedeutet, daß $u(\dot{z})\in U(\dot{z})$, $\dot{z}\in\dot{Z}$, ist. Schließlich sei $u-U$ durch $(u-U)(\dot{z}) := u(\dot{z}) - U(\dot{z})$ erklärt. Mit demselben Buchstaben K, der oben für den nichtnegativen Kegel in $\mathbb{R}^m$ eingeführt wurde, werden wir im weiteren auch dasjenige Element aus $L^m(\dot{Z})$ bezeichnen, welches jedem $\dot{z}\in\dot{Z}$ den Kegel K zuordnet.

Es sei $\{G^f,\ V^f(\dot{z})\}$, $\dot{z}\in\dot{Z}$, eine Lösung (2). Für jede Funktion g aus $F := \{f \mid \dot{Z}\to\dot{X}\}$ ist γ_f^g eine Abbildung von $\dot{Z}$ in $\mathbb{R}^m$, d. h. $\gamma_f^g\in M^m(\dot{Z})$. Mit $I_{\dot{x}}\in F$ bezeichnen wir diejenige Entscheidungsfunktion, die jedem $\dot{z}\in\dot{Z}$ ein und dieselbe Steuerung $\dot{x}\in\dot{X}$ zuordnet. Entsprechend (4) definieren wir die Funktion $\gamma_f^{I_{\dot{x}}}\in M^m(\dot{Z})$ durch

$$\gamma_f^{I_{\dot{x}}}(\dot{z}) := \Phi(\dot{z}) + \sum_{\dot{z}'\in\dot{Z}} P(\dot{z}' \mid \dot{z},\ \dot{x})\cdot V^f(\dot{z}')\ ,\quad \dot{z}\in\dot{Z}\ .$$

Mit dem folgenden Lemma wird gezeigt, daß die weiter unten betrachteten Mengen von minimalen Elementen von der leeren Menge verschieden sind. Der Beweis dieses Lemmas wird in Analogie zu Lemma 3.3. in [4] geführt.

Lemma 3: *Es seien $f\in F$ und $\{G^f,\ V^f(\dot{z})\}$, $\dot{z}\in\dot{Z}$, eine Lösung von* (2). *Dann gilt:*

a) Min $\{\bigcup_{\dot{x}\in\dot{X}} \gamma_f^{I_{\dot{x}}}\}\in L^m(\dot{Z})$;

b) $(\gamma_f^f - K)\cap$ Min $\{\bigcup_{\dot{x}\in\dot{X}} \gamma_f^{I_{\dot{x}}}\}\in L^m(\dot{Z})$.

Mit dem folgenden Satz wird ein Verfahren zur Strategieverbesserung angegeben.

Satz 2: *Die Folge (f_n), $f_n\in F$, sei wie folgt iterativ definiert:*

(1) *Man nehme eine beliebige Entscheidungsfunktion $f_0\in F$.*

(2) *Für jedes $n\geqq 0$ wähle man $f_{n+1}\in F$ derart, daß*

$$\gamma_{f_n}^{f_{n+1}} \in (\gamma_{f_n}^{f_n} - K)\cap \text{Min } \{\bigcup_{\dot{x}\in\dot{X}} \gamma_{f_n}^{I_{\dot{x}}}\}\ .$$

Dann gilt:

a) $G^{f_0}\geqq G^{f_1}\geqq \ldots \geqq G^{f_n}\geqq G^{f_{n+1}}\geqq \ldots$;

b) *wenn für ein N die Beziehung*

$$(\gamma_{f_N}^{f_N} - K) \cap \mathrm{Min}\,\{\bigcup_{\dot{x}\in\dot{X}} \gamma_{f_N}^{I_{\dot{x}}}\} = \{\gamma_{f_N}^{f_N}\} \tag{6}$$

erfüllt ist, so läßt sich die Strategie $\sigma=(f_N^\infty)$ in dem mit f_0 begonnenen Iterationsprozeß hinsichtlich der Folge von D-Kosten (G^{f_n}) nicht weiter verbessern.

Beweis: a) Da f_{n+1} für jedes n der Bedingung $\gamma_{f_n}^{f_{n+1}} \in (\gamma_{f_n}^{f_n} - K)$ genügt, folgt sofort $\gamma_{f_n}^{f_n} \geqq \gamma_{f_n}^{f_{n+1}}$. Nach Lemma 2 erhalten wir somit $G^{f_n} \geqq G^{f_{n+1}}$. Der Beweis von b) ist trivial. ■

Bemerkung: Die Existenz von f_{n+1} in (5) ergibt sich aus Lemma 3, b).

Aus der Bedingung (6) folgt i. a. für $m \geqq 2$ nicht, daß $\sigma=(f_N^\infty)$ eine D-optimale Strategie ist. Fordert man jedoch neben (6) zusätzlich, das $\gamma_{f_N}^{I_{\dot{x}}} \geqq \gamma_{f_N}^{f_N}$ für alle $\dot{x}\in\dot{X}$, so sind diese beiden Bedingungen äquivalent zu $\mathrm{Min}\,\{\bigcup_{\dot{x}\in\dot{X}} \gamma_{f_N}^{I_{\dot{x}}}\} = \{\gamma_{f_N}^{f_N}\}$ und es gilt

Satz 3: *Für die Strategie $\sigma=(f^\infty)$, $f\in F$, sei $\{G^f, V^f(\dot{z})\}$, $\dot{z}\in\dot{Z}$, eine Lösung von* (2). *Ist*

$$\mathrm{Min}\,\{\bigcup_{\dot{x}\in\dot{X}} \gamma_f^{I_{\dot{x}}}(\dot{z})\} = \{\gamma_f^f(\dot{z})\} \tag{7}$$

für alle $\dot{z}\in\dot{Z}$, so ist σ D-optimal.

Beweis: Wir zeigen, daß für jede von $\sigma=(f^\infty)$ verschiedene Strategie $\sigma'=(g^\infty)$ stets $G^g \geqq G^f$ ist, woraus die Behauptung folgt. Sei $g\in F$ mit $g \neq f$ beliebig. Ist $\dot{z}'$ ein Zustand mit $g(\dot{z}')=f(\dot{z}')$, so gilt offensichtlich $\gamma_f^g(\dot{z}')=\gamma_f^f(\dot{z}')$. Aus der Verschiedenheit der Funktionen f und g folgt, daß es wenigstens ein $\hat{\dot{z}}\in\dot{Z}$ mit $g(\hat{\dot{z}}) \neq f(\hat{\dot{z}})$ geben muß. Nach Voraussetzung (7) erhalten wir dann für $\hat{\dot{z}}$ die Beziehung $\gamma_f^{I_{\dot{x}}}(\hat{\dot{z}}) > \gamma_f^f(\hat{\dot{z}})$, die für alle $I_{\dot{x}}$ mit $\dot{x} \neq f(\hat{\dot{z}})$ erfüllt ist, also insbesondere auch für $\dot{x}=g(\hat{\dot{z}})$. Damit gilt $\gamma_f^g(\dot{z}) \geqq \gamma_f^f(\dot{z})$ für alle $\dot{z}\in\dot{Z}$, und nach Lemma 2 ist folglich $G^g \geqq G^f$. ■

4. Charakterisierung D-optimaler Strategien. Algorithmus zur Bestimmung der minimalen D-Kosten

Zur Charakterisierung optimaler Stretegien verwenden BLACKWELL [7] (für einen reellwertigen Entscheidungsprozeß) und FURUKAWA [4] (für ein vektorwertiges Modell) spezielle Optimalitätsgleichungen. In diesem Abschnitt führen wir eine Optimalitätsgleichung ein und zeigen, welche Beziehung zwischen dieser Gleichung und einer D-optimalen Strategie besteht.

Es seien $f\in F$ eine beliebige Entscheidungsfunktion und $\{G^f, V^f(\dot{z})\}$, $\dot{z}\in\dot{Z}$, eine Lösung von (2). Gilt für alle $\dot{z}\in\dot{Z}$

$$\gamma_f^f(\dot{z}) \in \mathrm{Min}\,\{\bigcup_{\dot{x}\in\dot{X}} \gamma_f^{I_{\dot{x}}}(\dot{z})\}\,, \tag{10}$$

so nennen wir $\gamma_f^f \in \operatorname{Min} \{\bigcup_{\dot{x} \in \dot{X}} \gamma_f^{I_{\dot{x}}}\}$ eine *Optimalitätsgleichung* und sagen, daß f der Optimalitätsgleichung genügt (bzw. eine Lösung der Optimalitätsgleichung ist). $\tilde{F}$ bezeichne die Menge aller dieser Entscheidungsfunktionen: $\tilde{F} := \{f \in F \mid \gamma_f^f \in \operatorname{Min} \{\bigcup_{\dot{x} \in \dot{X}} \gamma_f^{I_{\dot{x}}}\}\}$.

Definition 4: Eine Lösung f der Optimalitätsgleichung heißt *minimal*, wenn $\gamma_f^g \in \operatorname{Min} \{\bigcup_{\dot{x} \in \dot{X}} \gamma_f^{I_{\dot{x}}}\}$ für jede andere Lösung g der Optimalitätsgleichung gilt.

Mit $F^* := \{f \in \tilde{F} \mid \gamma_f^g \in \operatorname{Min} \{\bigcup_{\dot{x} \in \dot{X}} \gamma_f^{I_{\dot{x}}}\}$ für jedes $g \in \tilde{F}\}$ werde die Menge aller minimalen Lösungen der Optimalitätsgleichung bezeichnet. Ferner sei $F_f := \{g \in F \mid \gamma_f^f(\dot{z}) \leqq \gamma_f^g(\dot{z})$ für alle $\dot{z} \in \dot{Z}$ und $\gamma_f^f(\dot{z}) \prec \gamma_f^g(\dot{z})$ für wenigstens ein $\dot{z}\}$ mit $f \in F$ und $\hat{F} := = \bigcup_{f \in F^*} F_f$. Aus der Konstruktion der Mengen F_f und F^* ergibt sich, daß $F_f \cap F^* = \emptyset$ für jedes $f \in F^*$, und folglich ist $\hat{F} \cap F^* = \emptyset$. Offensichtlich kann F in der Form $F = F^* \cup \hat{F}$ dargestellt werden.

Mit dem folgenden Satz geben wir eine Charakterisierung für eine D-optimale Strategie an.

Satz 4: *Eine Strategie $\sigma = (f^{*\infty})$ ist genau dann D-optimal, wenn folgende Bedingungen erfüllt sind:*
a) *f^* ist eine minimale Lösung der Optimalitätsgleichung und*
b) *G^{f^*} ist ein minimales Element der Menge aller mittleren D-Kosten G^f mit $f \in F^*$.*

Beweis: Wir zeigen zunächst, daß a) und b) aus der D-Optimalität einer Strategie folgen. Sei $\sigma = (f^{*\infty})$, $f^* \in F$, eine D-optimale Strategie, und seien F^* und $\hat{F}$ wie oben definiert. Angenommen, a) ist falsch, d. h. $f^* \notin F^*$; aus $F = F^* \cup \cup \hat{F}$ und $F^* \cap \hat{F} = \emptyset$ ergibt sich dann, daß $f^* \in \hat{F}$ sein muß. Folglich existiert ein $f \in F^*$ mit $f^* \in F_f$. Das aber bedeutet, es gilt $\gamma_f^f(\dot{z}) \leqq \gamma_f^{f^*}(\dot{z})$ für alle $\dot{z} \in \dot{Z}$ und $\gamma_f^f(\dot{z}') \prec \gamma_f^{f^*}(\dot{z}')$ für wenigstens ein $\dot{z}' \in \dot{Z}$. Nach Lemma 2 erhalten wir somit $G^f \prec G^{f^*}$, was der D-Optimalität von σ widerspricht.

Ferner läßt sich für jedes $h \in \hat{F}$ ein $g \in F^*$ derart angeben, daß $h \in F_g$ ist. Wegen Lemma 2 ist dann $G^g \prec G^h$. Daraus kann die Beziehung

$$M^* := \operatorname{Min} \{G^f \mid f \in F\} = \operatorname{Min} \{G^f \mid f \in F^* \cup \hat{F}\} = \operatorname{Min} \{G^f \mid f \in F^*\} \tag{11}$$

abgeleitet werden. Nach Voraussetzung ist $\sigma = (f^{*\infty})$ D-optimal, d. h. $G^{f^*} \in M^*$. Aus (11) folgt somit b).

Umgekehrt, sei $f^* \in F^*$ und $G^{f^*} \in \operatorname{Min} \{G^f \mid f \in F^*\}$. Wegen (11) ist dann $G^{f^*} \in M^*$, d. h. $\sigma = (f^{*\infty})$ ist D-optimal. ■

Wir geben nun einen Algorithmus zur Bestimmung aller minimalen Lösungen der Optimalitätsgleichung an:

I. *Wähle $f_0 \in F$ beliebig.*

II. *Für jedes $n \geqq 0$ berechne man eine Lösung $\{G^{f_n}, V^{f_n}(\dot{z})\}$, $\dot{z} \in \dot{Z}$, von (2) und be-*

stimme $\gamma_{f_n}^{I_{\dot{x}}}(\dot{z})$ *für alle* $\dot{x}\in X$, $\dot{z}\in\dot{Z}$. *Wähle* $f_{n+1}\in F$ *derart, daß*

$$\gamma_{f_n}^{f_{n+1}}\in(\gamma_{f_n}^{f_n}-K)\cap\operatorname{Min}\ (\bigcup_{\dot{x}\in\dot{X}}\gamma_{f_n}^{I_{\dot{x}}}\}\ .$$

III. *Gilt für ein* N, *daß*

$$(\gamma_{f_N}^{f_N}-K)\cap\operatorname{Min}\ \{\bigcup_{\dot{x}\in\dot{X}}\gamma_{f_N}^{I_{\dot{x}}}\}=\{\gamma_{f_N}^{f_N}\}$$

(*d. h.*, f_N *genügt der Optimalitätsgleichung*), *so gehe man zu einem der beiden Fälle über:*

a) *Wenn* $\gamma_{f_N}^{f_N}\leqq\gamma_{f_N}^{I_{\dot{x}}}$ *für alle* $\dot{x}\in\dot{X}$, *so ist*

$$F^*=\{f_N\}\cup\{f\in F\mid\gamma_{f_N}^{f}(\dot{z})=\gamma_{f_N}^{f_N}(\dot{z})\quad\text{für alle}\quad\dot{z}\in\dot{Z}\}\ .$$

Stop.

b) *Existieren* $\dot{z}_0\in\dot{Z}$ *und* $\dot{x}_0\in\dot{X}$ *derart, daß*

$$\gamma_{f_N}^{I_{\dot{x}_0}}(\dot{z}_0)\in\operatorname{Min}\ \{\bigcup_{\dot{x}\in\dot{X}}\gamma_{f_N}^{I_{\dot{x}}}(\dot{z}_0)\}\quad\text{und}\quad\gamma_{f_N}^{I_{\dot{x}_0}}(\dot{z}_0)\neq\gamma_{f_N}^{f_N}(\dot{z}_0)\ ,\tag{12}$$

so werden aus F *alle* $f\neq f_N$ *entfernt, für die gilt:*

$$\gamma_{f_N}^{f_N}(\dot{z})\leqq\gamma_{f_N}^{f}(\dot{z})\quad\text{für alle}\quad\dot{z}\in\dot{Z}\quad\text{und}\quad\gamma_{f_N}^{f_N}(\dot{z})<\gamma_{f_N}^{f}(\dot{z})$$

für wenigstens ein $\dot{z}$.

Man definiere eine Funktion g *mit* $g\neq f_N$ *durch:*

$$g(\dot{z}_0)=\dot{x}_0\ ,\quad\text{wenn (12) erfüllt ist,}$$
$$g(\dot{z})=f_N(\dot{z})\quad\text{sonst.}$$

(*Die so konstruierte Strategie* $\sigma=(g^\infty)$ *nennen wir potentiell D-optimal bzgl. der Lösung* f_N *der Optimalitätsgleichung.*)

IV. *In* I *ersetze man* f_0 *durch* g *und gehe zu* II *und dann zu* III *über.*

V. *Die obigen Schritte werden solange ausgeführt, bis alle potentiellen D-optimalen Strategien bez. der Lösungen der Optimalitätsgleichung nach dem obigen Verfahren untersucht worden sind.*

VI. a) *Gibt es noch Funktionen* $h\in F$, *die beim Durchlaufen der Schritte* I *bis* V *nicht untersucht worden sind, so beginne man wieder mit* I *und ersetze* f_0 *durch* h.
b) *Wenn es keine derartigen Funktionen gibt, so gehe man zu* VII *über.*

VII. *Unter den in den obigen Schritten gefundenen Lösungen der Optimalitätsgleichung ermittle man sämtliche minimalen Lösungen. Damit ist* F^* *vollständig bestimmt.*

Der oben beschriebene Algorithmus ergibt sich aus den in Abschnitt 2 und 3 angeführten Aussagen. Wir bemerken zunächst, daß das im Satz 2 beschriebene Verfahren zur Strategieverbesserung gleichzeitig einen Algorithmus zur Bestimmung einer Lösung der Optimalitätsgleichung darstellt, denn aus (6) folgt $\gamma_{f_N}^{f_N}\in\operatorname{Min}\ \{\bigcup_{\dot{x}\in\dot{X}}\gamma_{f_N}^{I_{\dot{x}}}\}$. Bezüglich einer so gefundenen Lösung f_N der Optimalitätsgleichung lassen sich dann die unter III a) bzw. III b) genannten Fälle unterscheiden. Gilt $\gamma_{f_N}^{f_N} \leq \gamma_{f_N}^{I_{\dot{x}}}$ für alle $\dot{x} \in \dot{X}$, so folgt aus Satz 3, daß $\sigma = (f_N^\infty)$ D-optimal ist. Wegen Satz 4 ist dann $f_N \in F^*$. Jede Entscheidungsfunktion $f \in F$ mit $\gamma_{f_N}^{f_N}(\dot{z}) < \gamma_{f_N}^{f}(\dot{z})$ für wenigstens ein $\dot{z} \in \dot{Z}$ kann der Menge F^* nicht angehören, da $f \in F_{f_N}$ und $F_{f_N} \cap F^* = \emptyset$. Gibt es eine Funktion $g \in F$ mit $\gamma_{f_N}^{f_N}(\dot{z}) = \gamma_{f_N}^{g}(\dot{z})$ für alle $\dot{z} \in \dot{Z}$, so ist nach Lemma 2 $G^{f_N} = G^g$, d. h. $\sigma = (g^\infty)$ ist ebenfalls D-optimal, und folglich gilt $g \in F^*$. Damit ist F^* im Fall III a) durch $F^* = \{f_N\} \cup \{g \in F \mid \gamma_{f_N}^{f_N}(\dot{z}) = \gamma_{f_n}^{g}(\dot{z})$ für alle $\dot{z} \in \dot{Z}\}$ bestimmt. Gilt hingegen (12) in III b), so werden mit Hilfe des Algorithmus aus Satz 2 zunächst alle Lösungen der Optimalitätsgleichung ermittelt. Anschließend bestimmt man unter diesen die minimalen Lösungen.

Abschließend sei bemerkt, daß sich der Algorithmus zur Bestimmung der minimalen mittleren D-Kosten, d. h. der Menge M^*, aus den Schritten I bis VII des oben beschriebenen Verfahrens zur Ermittlung von F^* und der anschließenden Anwendung von (11) zusammensetzt.

Literatur

[1] Hoffmann, R.; R. Merker: Zur Theorie stochastischer zellularer Systeme. Z. elektr. Inform.- u. Energietechnik, Leipzig, **12** (1982) 28–37.

[2] Storm, J.: Über die Steuerung stochastischer zellularer Systeme. Wiss. Zeitschr. Pädagog. Hochschule Dresden, Mathem.-naturw. Reihe, **18** (1984) 29–37.

[3] Thomas, L. C.: Constrained Markov Decision Processes as Multi-objective Problems. *In:* Multi-objective Decision Making. The Institute of Mathematics and its Applications. Conference Series. Academic Press, London 1983, 79–92.

[4] Furukawa, M.: Vector-valued Markovian Decision Processes with Countable State Space. *In:* Recent Developments in Markov Decision Processes. Academic Press, London 1980, 205–223.

[5] Hinderer, K.: Foundations of Non-stationary Dynamic Programming with Discrete Time Parameter. Berlin 1970.

[6] Derman, C.: Denumerable State Markovian Decision Processes–Average Cost Criterion. Ann. of Math. Stat., **37** (1966) 1545–1553.

[7] Blackwell, D.: Discounted Dynamic Programming. Ann. Math. Stat., **36** (1965) 226–235.

Received April 1987, revised December 1987

J. Storm
Pädagogische Hochschule Dresden
Sektion Mathematik
Wigardstr. 17
Dresden
DDR-8060

optimization 19 (1988) 6, 905–906

Book Review

KARR, ALAN F.: **Point Processes and their Statistical Inference.** Vol. 2. Marcel Dekker New York, 1986, 490 S., ISBN 0-8247-7513-9.

Der Autor hielt die Zeit mit Recht für gekommen, die Statistik von Punktprozessen aus dem Zustand der Ansammlung diverser Methoden und Techniken in denjenigen einer Beschreibung von einem einheitlichen Zugang aus zu überführen.

Das Buch war ursprünglich als Forschungsmonographie gedacht, jedoch ist dem Verfasser zuzustimmen, wenn er es auch kennzeichnet als Basis für ein Seminar bzw. als ein Lehrbuch für Leser mit sehr gutem Grundwissen in Wahrscheinlichkeitstheorie einschließlich einer Einführung in diskrete Martingale (für zeitstetige Martingale bietet ein Anhang das Erforderliche) und normalem Wissen in mathematischer Statistik.

Es handelt sich um ein theoretisch anspruchsvolles, aber sehr gut geschriebenes und lesbares Werk, versehen mit kommentierenden und motivierenden Textteilen, ca. 300 Übungsaufgaben sowie Anwendungsbezügen. Konkrete Applikationen und Methoden der Datenanalysis gehören nicht zum Anliegen und Inhalt des Buches.

Wo möglich, werden zufällige Punktprozesse auf allgemeinen Räumen (lokal kompakt, separabel, HAUSDORFFsch) betrachtet, für spezielle Klassen wird nur die reelle Achse bzw. positive Halbachse zugelassen. Insbesondere werden folgende wichtige Klassen statistisch untersucht: Cluster Prozesse, Erneuerungs-, POISSON- und insbesondere COX-Prozesse, stationäre Punktprozesse, markierte Punktprozesse, unbegrenzt teilbare Punktprozesse, symmetrisch verteilte Punktprozesse, MARKOWsche Erneuerungsprozesse, regenerative Prozesse, leider jedoch nicht stochastische Prozesse mit eingebetteten nicht notwendig rekurrenten Punktprozessen.

Die Grundbegriffe der Punktprozeßtheorie einschließlich PALMscher Verteilungen auch für nicht notwendig stationäre Punktprozesse werden straff im kurzen Kap. 1 dargelegt. In Kap. 2 folgen, für Punktprozesse auf dem R_+, stochastische Integration und Intensität, Kompensator und damit Martingale.

Die allgemeinen Probleme und Vorgehensweisen für die Punktprozeßstatistik enthält Kap. 3 in guter Übersicht: Die verwendeten Hauptprinzipien für die Schätzung (Maximum Likelihood, Momentenmethode) und für Tests (Likelihood Quotienten) sowie spezielle strukturabhängige Methoden jeweils unter unterschiedlichen Beobachtungsstrategien (vollständige bzw. teilweise Beobachtung, eine oder mehrere Realisierungen). Es finden sich, dargelegt an Beispielen für Punktprozeßklassen, parametrische und nichtparametrische Schätzung ebenso wie parametrische bzw. nichtparametrische Tests quantitativer Hypothesen und Tests qualitativer Hypothesen. Weiterhin wird die Zustandsschätzung ausgeführt, d. h. die optimale Rekonstruktion nicht beobachteter Teile des Punktprozesses oder mit ihm verbundener Zufallsgrößen bzw. Prozesse, wobei der mittlere quadratische Fehler als Kriterium dient.

Empirische Maße und Funktionale sowie Schätzer für reduzierte PALMsche Verteilungen und verdünnte Punktprozesse untersucht Kap. 4.

Kap. 5 dagegen ist der allgemeinen Theorie der Martingal-Schätzmethoden für Punktprozesse auf dem R_+ mit stochastischer Intensität gewidmet. Als prinzipielle Vorzüge werden herausgestellt und im einzelnen ausgeführt: Die Möglichkeit, Konstruktionsregeln für Schätzer anzugeben, wenn andere Methoden versagen, effektive Methoden für die systematische Berechnung von Mittelwerten, Varianzen und Kovarianzen (als wesentlichster Beitrag für Anwendungen) und potentielle Techniken für die Ermittlung der asymptotischen Normalität von funktionswertigen Schätzern.

Für die Statistik von POISSONprozessen und rekurrenten Punktprozessen auf dem R_+ und stationären Punktprozessen in d-dimensionalen euklidischen Räumen enthalten die Kapitel 6 bis 9 die entsprechenden ausführlichen statistischen Verfahren und Ergebnisse.

Im Kap. 5 werden auch zwei Konzepte für die Untersuchung bedingter Größen in allgemeinen Räumen vorgestellt, in denen die für den Kompensator-Zugang auf dem R_+ wesentliche Eigenschaft der linearen Ordnung nicht vorliegt. Statt des natürlichen Zeitablaufes mit Vergangenheit, Stoppzeit, Prediction wird z. B. eine Außensicht-Sigmaalgebra eingeführt.

Das letzte, zehnte Kapitel umfaßt die Statistik für zufällige Prozesse (MARKOWsche, stationäre) und stationäre Felder, für die Beobachtungen durch einen Punktprozeßmechanismus, vornehmlich durch POISSONsche Probennahme, vorliegen.

Das Buch ist überzeugend und sehr exakt verfaßt, die Beweise sind überwiegend ausgeführt; fast 700 Literaturstellen sind angeführt. Der Verfasser verwendet beide Zugänge zu Punktprozessen, den sog. Verteilungszugang (zufälliges Zählmaß) bzw. den Martingal-Zugang, je nach betrachtetem Grundraum und Zweckmäßigkeit, die jeweils herausgestellt wird.

Erstmalig liegt ein Werk dieser Art vor, und es empfiehlt sich durch Gegenstand, Umfang und Art der Bearbeitung und Darstellung sogleich als ein Standardwerk in Forschung und Ausbildung für anspruchsvolle Aussagen der Punktprozeßstatistik.

D. KÖNIG

Acknowledgement

The editor likes to express his deep gratitude to all the referees who helped the editorial board to decide about so many papers submitted for publication in our journal. Special thanks from the editorial staff to Dr. rer. nat. R. Reinhardt for his dependable and valuable support during the last twelfe years.

The following referees were active during the last two years:

J. Bair (Liège)
E. J. Balder (Utrecht)
H. Bandemer (Freiberg)
K. Beer (Karl-Marx-Stadt)
F. Behringer (München)
F. Beichelt (Mittweida)
D. Bierlein (Regensburg)
M. Borchardt (Ilmenau)
B. Brosowski (Frankfurt/M.)
R. E. Burkard (Graz)
U. Dave (London)
V. Demyanov (Leningrad)
G. Dewess (Leipzig)
S. Dietze (Dresden)
S. Dolecki (Limoges)
A. L. Dontchev (Sofia)
R. Elster (Merseburg)
B. Engelmann (Leipzig)
H. Erfurth (Merseburg)
Y. G. Evtushenko (Moscow)
B. Freyer (Berlin)
T. Gal (Hagen)
C. Gerth (Merseburg)
F. Giannessi (Pisa)
H. Gillert (Dresden)
E. Girlich (Jena)
H.-J. Girlich (Leipzig)
A. Göpfert (Merseburg)
G. Griessbach (Jena)
H.-D. Gronau (Greifswald)
T. Hanschke (Mainz)
J. Hansohm (Augsburg)
J. Harant (Ilmenau)
H. Hartwig (Leipzig)
L. Hempel (Weimar)
R. Hettich (Trier)
K. Hinderer (Karlsruhe)
A. Hoffmann (Ilmenau)
R. Horst (Trier)
J. Jahn (Erlangen)
H. Th. Jongen (Aachen)
J. Käschel (Karl-Marx-Stadt)
P. Kall (Zurich)
A. A. Kaplan (Novosibirsk)
D. Klatte (Halle)
H. Klemm (Jena)
R. Klötzler (Leipzig)
P. Köchel (Karl-Marx-Stadt)
D. König (Freiberg)
M. Kolonko (Hildesheim)
K. O. Kortanek (Iowa City)
W. Krabs (Darmstadt)
H.-U. Küenle (Leipzig)
B. Kummer (Berlin)
C. Lemarechal (Le Chesnay)
P. O. Lindberg (Stockholm)
F. A. Lootsma (Delft)
B. Luderer (Karl-Marx-Stadt)
H.-J. Lüthi (Zurich)
G. Maess (Rostock)
K. Malanowski (Warsaw)
O. L. Mangasarian (Madison)
L. Martein (Pisa)
J.-E. Martinez-Legaz (Barcelona)
P. Mazzoleni (Verona)
R. Mennicken (Regensburg)
P. H. Müller (Dresden)
G. Nägler (Leipzig)
S. V. Nagaev (Novosibirsk)
J. Naumann (Berlin)
R. Nehse (Ilmenau)
V. Nollau (Dresden)
W. Oettli (Mannheim)
C. Olech (Warsaw)
D. Pallaschke (Karlsruhe)
J.-P. Penot (Pau)
J. Piehler (Merseburg)
B. N. Pshenichnyi (Kiev)
F. J. Radermacher (Ulm)
C. Richter (Köthen)
K. Richter (Karl-Marx-Stadt)
J. Rohn (Prague)
H. Rudolph (Leipzig)
S. Schaible (Riverside)
G. Scheithauer (Dresden)
W. Schirotzek (Dresden)
K. Schittkowski (Bayreuth)
J. W. Schmidt (Dresden)

V. Schmidt (Freiberg)
W. Schneiderheinze (Magdeburg)
M. Schoch (Freiberg)
K. Schönefeld (Dresden)
H.-J. Sebastian (Leipzig)
J. Seeländer (Merseburg)
E. Spedicato (Bergamo)
P. Starke (Berlin)
R. E. Steuer (Athens)
J. Stoer (Würzburg)
M. M. Sysło (Wroclaw)
F. Tardella (Pisa)
H.-J. Terno (Dresden)
R. Theodorescu (Quebec)
J. Thierfelder (Ilmenau)
R. Tichatschke (Güstrow)
N. V. Tretjakov (Moscow)
F. Tröltzsch (Karl-Marx-Stadt)
Hoang Tuy (Hanoi)
M. Vlach (Prague)
H.-J. Walther (Ilmenau)
P. Weidner (Halle)
P. L. Yu (Lawrence)
H.-J. Zimmermann (Aachen)
K. Zimmermann (Prague)
S. Zionts (Buffalo)
S. Zlobec (Montreal)
J. Zowe (Bayreuth)

URSULA SCHULZE

Mehrphasenregression

Stabilitätsprüfung – Schätzung – Hypothesenprüfung

1987. 178 Seiten – 40 Abbildungen – 15 Tabellen – 14,5 cm × 21,5 cm
DDR 24,– M, Ausland 30,– DM
ISBN 3-05-500224-5
Bestell-Nr. 763 622 5
Bestellwort: Schulze, Verfahren 6971

Die Autorin behandelt Verfahren zur Stabilitätsprüfung eines Regressionsmodells (kumulative Summentests), zur Schätzung der Umschlagspunkte und Phasenparameter und zur Prüfung von Hypothesen über die Phasenparameter und Umschlagspunkte in Modellen mit gegebener Phasenanzahl. Sie wendet sich damit an mathematisch interessierte Biologen, Mediziner, Pharmazeuten, Chemiker, Landwirtschaftswissenschaftler, Psychologen, Techniker und andere Wissenschaftler, die sich mit der Auswertung von Daten befassen.
Das Buch ist leicht zu lesen und mit zahlreichen Beispielen aus Biologie und Medizin versehen, durch die der Leser gut den Weg von den mathematischen Sätzen hin zur Realisierung der Verfahren durch Computerprogramme verfolgen kann. Es erfordert in der Regel nur Grundkenntnisse in mathematischer Statistik.

optimization

a journal of mathematical programming and operations research

Volume 19 1988 Number 6

Contents

The Annual Index of Volume **19** (1988) is enclosed in this Number.

optimization is indexed in: Comp. Inf. Syst. Abstr., Int. Abstr. Operat. Res., J. Cont. Quant. Meth., Math. Rev., Ref. Ž., Zbl. Math.

Bei Fragen zur Produktsicherheit wenden Sie sich bitte an:
If you have any questions regarding product safety,
please contact:

Walter de Gruyter GmbH
Genthiner Straße 13
10785 Berlin
productsafety@degruyterbrill.com